Praise for

THE KILLER ACROSS THE TABLE

"A must read for everyone wanting to understand the minds of depraved serial killers. Once again John Douglas walks into rooms most of us would shun in our worst nightmares and comes back with remarkable insight into what type of person perpetrates such horrible acts."

—NEW YORK JOURNAL OF BOOKS

"This fascinating and haunting expert account helps us understand why the most shocking homicides occur. . . . [This] will be of particularly urgent interest."

—BOOKLIST *(STARRED REVIEW)*

"An eye-opening and chilling look into the minds of brutal murderers. . . . Douglas succeeds in taking readers inside his own mind as he worked to understand 'four killers [he] confronted.'. . . Separating fact from fiction, Douglas adds to lay readers' understanding of human evil."

—*PUBLISHERS WEEKLY*

"These sharply written, detailed investigations keep to the facts, and that is where their worth and horror lie. Go beyond the fading headlines to deep inside prison walls, across small tables in windowless rooms and face-to-face with the men whose crimes made them monsters."

—*BOOKPAGE*

"As a true-crime writer, Douglas has a perspective and depth of experience that few share, and he and his writing partner don't flinch from exploring the complexities of the psyches of his subjects."

—*COLUMBUS DISPATCH*

THE KILLER ACROSS THE TABLE

ALSO BY JOHN DOUGLAS AND MARK OLSHAKER

Mindhunter

Journey into Darkness

Unabomber

Obsession

The Anatomy of Motive

The Cases That Haunt Us

Broken Wings

Law & Disorder

THE KILLER ACROSS THE TABLE

UNLOCKING THE SECRETS OF SERIAL KILLERS AND PREDATORS WITH THE FBI'S ORIGINAL MINDHUNTER

JOHN DOUGLAS AND
MARK OLSHAKER

DEY ST.
An Imprint of WILLIAM MORROW

 For information, address HarperCollins Publishers, 195 Broadway, New York, NY 10007.

HarperCollins books may be purchased for educational, business, or sales promotional use. For information, please email the Special Markets Department at SPsales@harpercollins.com.

A hardcover edition of this book was published in 2019 by Dey Street, an imprint of William Morrow.

FIRST DEY STREET PAPERBACK EDITION PUBLISHED 2020.

Designed by Renata De Oliveira

Abstract paint background image on title page and part titles by Eky Studio/shutterstock
Splattered bloodstain background image on table of contents by oriontrail/shutterstock

Library of Congress Cataloging-in-Publication Data has been applied for.

ISBN 978-0-06-291064-6

20 21 22 23 24 DIX/LSC 10 9 8 7 6 5 4 3 2 1

To the memory of Joan Angela D'Alessandro and in honor of Rosemarie D'Alessandro and all of the others who, through their inspiration, courage, and determination, strive for justice and safety for all children, this book is dedicated with love and admiration

AUTHORS' NOTE

The opinions expressed in this book belong to the authors alone and do not reflect those of the FBI or any other organization.

For photos related to the cases in this book and other information about John Douglas's career, the authors and their work, please visit www.mindhuntersinc.com.

IN A SMALL ROOM IN THE BIG HOUSE

Here, it is not so much *Who done it?*, but *Why?*

And in the end, if we have discovered the *Why?* and add in *How?*, we will also come to understand the *Who?* Because *Why? + How? = Who.*

The aim is not to be a friend. The aim is not to be a foe. The aim is to get to the truth.

It is a verbal and mental chess match without any game pieces; a sparring session without body contact; an endurance contest in which each side will seek out and exploit the other's weaknesses and insecurities.

We sit across a small table from each other in a dimly lit room whose cinder-block walls are painted a pale bluish gray. The only window is in the locked steel door, and it is small and reinforced with wire mesh. A uniformed guard peers through from the other side, making sure everything remains in order.

In a maximum security prison, nothing is considered more important.

We have been at this for two hours already and finally the moment is ripe. "I want to know in your own words what it was like twenty-five years ago," I say. "How did this all happen to get you here? That girl—Joan—did you know her?"

"Well, I'd seen her in the neighborhood," he replies. His affect is calm and his tone is even.

"Let's go back to the moment she came to the door. Tell me what happened, step by step, from that point on."

It is almost like hypnosis. The room is silent, and I watch him transform in front of me. Even his physical appearance seems to change before my eyes. His eyes are unfocused and he looks beyond me to stare at the vacant wall. He is moving back to another time and another place; to the one story of himself that has never left his mind.

The room is very cold, and even though I wear a suit, I struggle to keep myself from shivering. But as he recounts the story I have asked for, he has begun to perspire. His breathing grows heavier and more audible. Soon his shirt is drenched with sweat, and underneath, the muscles of his chest tremble.

He relates the entire story in this manner, not looking at me; almost talking to himself. He is in the zone, in that time and that place, thinking now what he was thinking then.

For a moment, he turns back to face me. He looks me square in the eyes as he says, "John, when I heard the knock and looked up through the screen door and saw who was there, I knew I was going to kill her."

INTRODUCTION

LEARNING FROM THE EXPERTS

This is a book about the way violent predators think—the bedrock of my twenty-five years as an FBI special agent, behavioral profiler, and criminal investigative analyst, as well as the work I have done since my retirement from the bureau.

But it's really a book about conversations I had. After all, conversations are where it all began for me, conversations in which I learned how to use what a predatory criminal was thinking to help local law enforcement officials to catch him and bring him to justice. For me, that was the beginning of behavioral profiling.

I started interviewing incarcerated violent offenders out of what I considered personal and institutional necessity, but in many ways, it began with a desire to understand the underlying motivations behind criminals. Like most new FBI special agents, I was assigned as a street agent. My first posting was in Detroit. Right from the beginning, I was interested in why people committed their crimes—not only *that* they committed crimes at all, but *why* they committed the particular crimes they did.

Detroit was a tough city, and while I was there they were racking up as many as five bank robberies a day. Robbing a bank backed by the Federal Deposit Insurance Corporation is a federal crime, so the bureau had jurisdiction, and many new agents were assigned to investigate these cases in addition to their other duties. As soon as we would apprehend a suspect and read him his Miranda rights, often in the back of a bureau car or police cruiser, I would pepper him with questions. Why rob a bank where the security is tight, and everything is recorded on tape, rather than a store that does a large cash business? Why this particular bank branch? Why this particular day and time? Was the robbery planned or spontaneous? Did you surveil the bank first and/or make a practice run inside? I began mentally cataloging the responses and developing informal "profiles" (though we didn't use that term yet) of bank robber types. I started seeing the differences between planned and unplanned crimes and organized and disorganized ones.

We got to the point where we could begin predicting which bank locations were most likely to be hit and when. In areas with a lot of building going on, for instance, we learned that Friday late morning was a likely time to hit the bank, because it would have a lot of cash on hand to handle the construction workers' paychecks. We used this kind of intelligence to harden certain targets and be waiting at others if we thought we had a reasonable chance of catching robbers in the act.

During my second bureau posting in Milwaukee, I was sent to the new, modern FBI Academy in Quantico, Virginia, for a two-week in-service course in hostage negotiation. It was taught by Special Agents Howard Teten and Patrick Mullany, the original champions of behavioral science in the bureau. Their main course was called Applied Criminology. It was an attempt to bring the academic discipline of abnormal psychology into crime analysis and the training of new agents. Mullany saw hostage negotiation as the first practical use of the applied psychology program. This was a new wave in the battle against a new era of crime that involved airplane skyjackings and bank robberies with hostages, such as the 1972 Brooklyn bank heist that inspired the Al Pacino film *Dog Day Afternoon*. It was easy to see how having some idea of what was going on in the hostage-taker's head would be of great benefit to the negotiator and ultimately save lives. I was one of about fifty special agents in this class, taught for the first time and a daring experiment in FBI training. Legendary director J. Edgar Hoover had died just three years before, and his lengthened shadow still hung over the bureau.

Even in his declining years, Hoover maintained an iron grip on the agency he had essentially created. His hard-nosed, hard-ass approach to investigation echoed the old *Dragnet* TV show line: *Just the facts, ma'am.* Everything had to be measurable and quantifiable—how many arrests, how many convictions, how many cases closed. He never would have embraced anything as impressionistic, inductive, and "touchy-feely" as behavioral science. In fact, he would have considered it a contradiction in terms.

While attending the hostage negotiation course at the FBI Academy, my name got around the Behavioral Science Unit, and before I left to go back to Milwaukee, I was offered positions in both the Education and Behavioral Science Units as my next posting. Even though our unit was called Behavioral Science, the primary responsibility of its nine agents was teaching. Courses included Applied Criminal Psychology, Hostage Negotiation, Practical Police Problems, Police Stress Management, and Sex Crimes, which was later changed by my great colleague Roy Hazelwood to Interpersonal Violence.

Though the academy's "three-legged stool" model of teaching, research, and consultation was beginning to take shape, any case consultation that star agents like Teten provided was strictly informal and not part of any organized program. The focus of these forty hours of classroom instruction was supposed to be on the issue of most concern to criminal investigators: *Motive*. Why do offenders do the things they do, in what ways do they do them, and how can understanding this help catch them? The problem with this approach was that most of the content still came from the academic sphere, which became evident whenever senior law enforcement personnel going through the National Academy program had more firsthand case experience than the instructors.

No one was more vulnerable in this area than the youngest instructor on the team: me. Here I was standing in front of a classroom full of seasoned detectives and officers, most of them a lot older than me. And I was supposed to teach them what was going on in the criminal mind, something they'd actually be able to use to help clear cases. Most of my firsthand experience had come from working with experienced cops and homicide detectives in Detroit and Milwaukee, so it seemed presumptuous of me to be telling men like that their own business.

There was a dawning realization among many of us that what was applicable to the psychiatric and mental health community had only limited relevance to law enforcement.

Still, I started getting the same types of requests Teten got. In class or during breaks, or even in the evenings, officers and detectives would come up and ask for pointers or advice on their active cases. If I was teaching a case similar in some way to one they were working on, they would figure I could help them solve it. They saw me as the authoritative voice of the Federal Bureau of Investigation. But was I? There had to be a more practical way to amass useful data and case studies that would give me the confidence to feel I really knew what I was talking about.

As the guy closest to me in age, Robert Ressler was tapped to help me break into the academy culture and feel comfortable with teaching. About eight years older than I was, Bob was a new instructor who built on what Teten and Mullany had done, aiming to bring the discipline of behavioral analysis closer to something that could be of value to police departments and criminal investigators. The most efficient way to give a new instructor some concentrated experience was through what we called road schools. Instructors from Quantico would spend a week teaching a selective piece, sort of a highlights version, of the National Academy curriculum to a police department or law enforcement agency that had requested it, then move on to another for a second week before returning home with the memories of interchangeable hotel rooms and a suitcase full of dirty laundry. So Bob and I hit the road together.

One morning early in 1978, Bob and I were driving out of Sacramento, California, the site of our latest road school. I commented that most of the criminals we were teaching about were still around, we could easily find out where they were, and they weren't going anywhere. Why not see if we could meet and talk with some of them, find out what a crime was like through *their* eyes, get them to recall and tell us *why* they did what they did and *what* was going on in their minds when they did it. I figured it didn't hurt to try, and some of them might be bored enough with their prison routines that they'd welcome a chance to talk about themselves.

Very little research was available relative to interviewing prison inmates, and what there was pertained specifically to convictions, probation and parole, and rehabilitation. However, the record seemed to indicate that violent and narcissistic inmates, on the whole, were incorrigible—meaning they were not able to be controlled, improved, or reformed. By talking to them, we hoped to learn if this was indeed the case.

Bob was initially skeptical, but willing to go along with this crazy idea. He'd served in the army, and between the army and the bureau, he'd had enough experience with bureaucracies that his mantra was "It's better to ask for forgiveness than permission." We would show up unannounced. In those days FBI credentials could get us into prisons without prior permission. If we had said in advance that we were coming, there was a danger word would leak out into the yard. And if an inmate was known to be planning to speak with a couple of feds, the rest of the prison population might think he was a snitch.

As we embarked on this project, we had some preconceived ideas of what we would encounter during these interviews. Among them:

- All would claim they were innocent.
- They would blame their convictions on poor legal representation.
- They would not willingly talk to law enforcement personnel.
- Sex offenders would come across as sex-obsessed.
- If there had been capital punishment in the state in which the murder was committed, they would not have killed their victims.
- They would project the blame onto the victims.
- They all came from dysfunctional family backgrounds.
- They knew the difference between right and wrong and the nature of the consequences of their actions.
- They were not mentally ill or insane.
- Serial murderers and rapists would tend to be highly intelligent.
- All pedophiles are child molesters.
- All child molesters are pedophiles.
- Serial killers are made, rather than born that way.

As we'll see in the following pages, some of these assumptions proved correct, while others were way off the mark.

Surprisingly, the overwhelming majority of the men we sought out did agree to talk to us. They had various reasons. Some felt co-operating with the FBI would look good on their records, and we did nothing to dissuade that assumption. Others might simply have been intimidated. Many inmates, particularly the more violent types, don't get many visitors, so it was a way to relieve the boredom, talk to some-one from the outside, and spend a couple of hours outside their cells. There were some who were just so cocksure of their ability to con everyone that they looked at the interview as a potential game.

In the end, what started with a simple idea while driving out of Sacramento—conversations with killers—became a project that would change the careers and lives of both Bob and me and the special agents who eventually joined the team, and add a new dimension to the FBI's crime-fighting arsenal. Before we were done with our initial phase of interviews, we had studied and talked to, among others: shoe fetishist and strangler Jerome Brudos in Oregon, who liked to dress his dead victims in high heels from his extensive wardrobe of women's clothing; Monte Rissell, who raped and murdered five women as a teenager in Alexandria, Virginia; and David Berkowitz, the .44 Caliber Killer and Son of Sam, who terrorized New York City in 1976 and 1977.

Over the years, my profilers at Quantico and I would interview many other violent and serial predators, including Ted Bundy, the prolific killer of young women, and Gary Heidnik, who imprisoned, tortured, and killed women in the basement pit of his house in Philadelphia. Both of these guys provided character traits for novelist Thomas Harris in *The Silence of the Lambs*, as did Ed Gein, the Wisconsin recluse who killed women so he could use their skins, whom I interviewed at the Mendota Mental Health Institute in Madison. He was also famously the model for Norman Bates in Robert Bloch's novel *Psycho*, the basis for the classic Alfred Hitchcock film. Unfortunately, Gein's age and mental illness resulted in such rambling, disordered thought patterns that the interview wasn't productive. He did, however, still enjoy working in leather crafts, making wallets and belts.

What eventually emerged was a set of rigorous interview methods that allowed us to start correlating the crime to what was actually in the criminal's mind at the time. For the first time, we had a way of linking and understanding what was going on in an offender's mind with evidence he left at the crime scene and what he said to the victim if she or he was alive, or what was done to the body, both ante- and postmortem. As we have often stated, it helped us begin to answer the age-old question "What type of person could do such a thing?"

By the time we had completed our initial round of interviews, we knew what type of person could do such a thing, and three words seemed to characterize the motivations of every one of our offenders: *Manipulation. Domination. Control.*

The conversations were the starting point for everything that came after. All the knowledge we gathered, the conclusions we drew, the *Sexual Homicide* book that came out of our research and the *Crime Classification Manual* that we created, the killers we helped catch and prosecute—all of it began by sitting across from killers and asking them about their lives with an aim to understanding what drove them to take another life, or in some cases, many lives. It was all possible because of the attention we paid to this previously untapped group of instructors: the criminals themselves.

We are going to take an in-depth look at four killers I confronted across the table after I had left the bureau, using the same techniques we had developed during our extensive study. The killers themselves are all different, each with his own techniques, motivations, and psychic makeup. They range from a single victim to close to a hundred, and I have learned from all of them. The contrasts between them are intriguing and compelling. But so are the similarities. They are all predators, and all grew up without forming trusting bonds with other human beings during their formative years. And they are all prime exhibits in one of the central debates of behavioral science: nature versus nurture, whether killers are born or made.

In my FBI unit, we operated under the equation *Why? + How? = Who.* When we interview convicted offenders, we can reverse-engineer that process. We know the *Who* and we know the *What*. By combining those, we discover the all-important *How?* and *Why?*

I

THE BLOOD OF THE LAMB

1

LITTLE GIRL LOST

It was just after the July Fourth holiday in 1998 when I took the Amtrak train up north to call on a new potential "instructor." His name was Joseph McGowan, and he had been a high school chemistry teacher with a master's degree. But rather than any formal academic title, he was now officially known as Inmate No. 55722 at his longtime place of residence, the New Jersey State Prison at Trenton.

The reason for his incarceration: the sexual assault, strangulation, and blunt force murder of a seven-year-old girl who had come to his house to deliver two boxes of Girl Scout cookies twenty-five years earlier.

As the train wound its way north, I prepared. Preparation when talking to a killer is always important, but no more so than now—after all, this conversation would have consequences far beyond the informational or academic. I'd been called in by the New Jersey parole board to help determine whether McGowan, who'd already been denied parole twice, should be released back into society.

At the time, the chairman of the New Jersey State Parole Board was an attorney named Andrew Consovoy. He had joined the parole board in 1989, and as McGowan's case was coming up for a third time, he had just been appointed chairman. Consovoy had read our book *Mindhunter* after hearing me one night on the radio and recommended it to the parole board's executive director Robert Egles.

"One of the things I realized from reading it and your other books was that you had to have all of the information going in," Consovoy related years later. "You had to find out who these people were. They didn't start to exist the day they came to jail."

Based on this perspective, he formulated a special investigations unit operating under the parole board. It consisted of two former police officers and a researcher, and its function was to look deeply into the questionable parole cases and give board members as much information as possible about the applicant on which to formulate a decision. They asked me to consult on the McGowan case.

Consovoy and Egles picked me up at the train station and took me to my hotel in Lambertville, a picturesque town on the Delaware River. There Egles handed over copies of everything in the case file.

The three of us went out to dinner that evening and talked generally about the work I did, but we stayed away from the specifics of the case. All they had told me was that the subject had killed a seven-year-old female child and they wanted to know whether he remained dangerous.

After dinner, they dropped me off back at the hotel, where I opened the case files and began several hours of review. My role was to see what I could determine about McGowan's state of mind—then and now. Did he know the nature and consequences of his crime? Did he know basic right from wrong? Did he care about what he had done? Did he have any remorse?

What would be his demeanor during the interview? Would he recall specific details about the crime? If released from prison, where did he intend to live and what did he intend to do? How would he earn a living?

My one cardinal rule of prison interviews is never to go into the encounter unprepared. I also made a practice of not going in with notes, because that could create an artificial distance or filter between the subject and me when the time came to really bore in and search for the deepest layer of his psyche.

I didn't know what I was going to get out of this interview, but I figured it would be illuminating. Because as I said at the beginning, every time I talked to "the experts," I learned something valuable. And one of the things to be determined was just what kind of expert Joseph McGowan would turn out to be.

I sifted through the case files, reexamining the evidence and organizing my thoughts for the next day's interview.

As I did, a grim story unfolded.

ABOUT 2:45 ON THE AFTERNOON OF APRIL 19, 1973, WHICH HER MOTHER ROSEMARIE would always remember was Holy Thursday, Joan Angela D'Alessandro noticed a car pulling into the first driveway on the right, on St. Nicholas Avenue, which intersected with Florence Street, where she lived. Joan and her older sister, Marie, had managed to sell Girl Scout cookies to just about everyone in a four-block area in their quiet Hillsdale, New Jersey, neighborhood. At that time, kids of that age going out by themselves to sell the cookies was a normal activity. Since they went to a Catholic school, the D'Alessandro girls had the day off for the religious holiday and spent part of it delivering orders. The people who lived in the house on the corner were the last customers they had to deliver to and then the cookie orders would be complete. Typically, Joan wanted to get the job done.

She was seven years old, a four-foot-three-inch bundle of playful energy and charm—a pretty, proud, and enthusiastic Brownie. In fact, she was enthusiastic about everything: school, ballet, drawing, dogs, dolls, friends, and flowers. Her second-grade teacher called her a "social butterfly," who naturally attracted people around her. Her favorite music was the "Ode to Joy" from Beethoven's Ninth Symphony. She was the youngest of three children, born close together. Frank, known as Frankie, was nine and Marie was eight. They were more serious, Rosemarie recalls. Joan was more happy-go-lucky.

"Joan was empathetic right from the beginning. She was always concerned with other people's feelings and hurts. And she had a natural spunkiness about her."

There is hardly a photograph of her at this age in which she isn't smiling: Joan in her Brownie uniform with its orange tie and beanie, hands clasped in front of her and long auburn hair symmetrically draping her shoulders; Joan in her black leotard and white tights, hair in a ponytail, arms outstretched to one side, demonstrating a ballet move; Joan in her navy blue plaid jumper, white blouse and red bow tie, as if she'd just turned to camera, bangs brushing her forehead and hair cascading around her adorable face; Joan sitting on her heels in a light blue party dress, hair pinned up, meticulously adjusting the bouquet in the hand of her Miss America Barbie doll. All of them represent different Joan personas. The two commonalities among them are the angelic smile and the innocent magic in her blue eyes.

A friend of Frankie's said, "She was so down-to-earth. I would have married her!"

Her Italian-speaking grandfather adored her and used to say, "*E così libera!*" She is so free! She had a hearty laugh and Rosemarie envisioned her acting in plays as she got older. She was going to be taking piano lessons after her eighth birthday.

This afternoon, she was outside playing by herself. Frankie had gone to play at his friend's house in the neighborhood and Marie was at a softball game.

Suddenly she raced back inside and said to Rosemarie, "I saw the new car. I'm going to take the cookies over there." She grabbed her Girl Scout carrying case lying in the foyer with the two boxes of cookies inside.

"Bye, Mommy. I'll be right back," she called out as she bounded out the front door. It hadn't even closed since she'd come running inside. Rosemarie remembers her ponytail bobbing up and down, held in place by an elastic band with two little light blue plastic balls on the ends, as Joan skipped down the front steps to the driveway and out onto the street. It all went by in a blur.

About ten minutes later the next-door neighbor, as she told Rosemarie afterward, heard the insistent barking of her dog, Boozer. Joan loved walking and playing with Boozer, and Boozer loved her.

When Joan didn't come back right away, Rosemarie didn't think much about it. She had probably gone to her friend Tamara's house on the corner of St. Nicholas Avenue and Vincent Street. It was that kind of neighborhood, where you could go in and out of the houses of people you knew. The social butterfly could always find someone to hang out with or something to do. About 4:45, when the music teacher came for Marie's piano lesson, Rosemarie started to get worried. She didn't want to convey it to the children, so she tried to hold herself together. After all, it was a safe neighborhood, with an FBI agent and a minister living nearby.

She started making phone calls. Joan wasn't at any of the houses she called, and no one had seen her.

Her husband, Frank D'Alessandro, got home about ten minutes to six and Rosemarie told him Joan was missing. Frank was a computer systems analyst, methodical and taciturn by nature. Rosemarie could see instantly how worried and tense he was, but as usual, he held it all in. Rosemarie said, "We have to call the police." Frank agreed and made the call. Then he went out with Frankie and Marie to drive around the neighborhood looking for Joan. They covered the entire area.

When they returned without having spotted her or found anyone who had seen her, Rosemarie decided to go out herself. Frank didn't want to come. She remembered that as Joan ran out, she'd said something about collecting on the last of her cookie orders because she'd seen "the new car" on St. Nicholas Avenue. The car belonged to the McGowan house. Joseph McGowan taught chemistry at Tappan Zee High, just over the state line in Orangeburg, New York. The house was owned by his mother, Genevieve McGowan, and he lived there with her and Genevieve's mother—his grandmother. The public schools had class that day, so this would have been about the right time for him to be coming home.

Reluctantly, so she wouldn't be alone, Rosemarie took Frankie with her and together they walked up Florence Street and turned onto St. Nicholas Avenue. It was ten minutes to seven. The McGowan house, a redbrick-and-beige-siding bi-level with a driveway and two-car garage in the left front, was the first one on the right, occupying the corner lot.

They both climbed the five front steps and she rang the bell. She told Frankie to stay on the landing.

Mr. McGowan answered the door. He seemed as if he'd just come out of the shower. He was holding a thin cigar that Rosemarie didn't notice at first. He was a twenty-seven-year-old bachelor. Rosemarie didn't know him, but "my children said he was very nice."

Rosemarie stepped into the foyer; she wanted to stand exactly where she knew Joan had recently stood. She was already starting to get an eerie feeling. She introduced herself. "Have you seen my daughter Joan?" she asked. "She came here to deliver cookies."

"No, I never saw her," he replied.

He spoke in a casual, matter-of-fact manner. And it was at that moment that Rosemarie D'Alessandro felt everything go cold.

"After standing in the foyer for a couple of minutes, I noticed a long fire truck parked in front of his house," she said. "We had called the police, and when I saw that they were responding this way, it just all came to me at that moment and I knew my life would never be the same."

She was almost immediately struck by McGowan's reaction—or rather, his lack of reaction. "As I was standing there in the foyer with him, tears were welling in my eyes. And he looked at me like he absolutely didn't have one ounce of feeling. And what he did at that moment when he saw my tears, he walked up the steps to the upper floor, and he stayed right there facing me, holding his slim cigar, and waited for me to leave.

"Walking back to my house I knew that he knew what had happened to Joan."

After the police arrived and spoke to Rosemarie and Frank, a neighborhood search for Joan was organized. Boy Scouts volunteered. As did Joseph McGowan. Hundreds of people turned out and organized themselves into small teams, checking every house, backyard, trash and garbage cans, woods, and park in Hillsdale and the surrounding towns. The police brought in bloodhounds to aid in the search. Several people climbed on the fire truck Rosemarie had seen parked. One of them was Joan's seven-year-old "boyfriend," Rich. They rode out to the reservoir near Woodcliff Lake.

At about 9:20, a priest from St. John the Baptist Church arrived at the house with a state trooper and a German shepherd. Rosemarie led the K-9 team to the clothes hamper so the dog could sniff Joan's panties, then they went out into the neighborhood. Rosemarie had the overwhelming sense that the dog understood what had happened and "felt" deeply for her and Joan. With an obvious sense of mission, he cased the area up to and around the McGowan's house, and went to the front door and the garage door.

But nothing turned up anywhere.

Word of the missing girl and the impromptu search spread quickly. Newspaper and television reporters swarmed the neighborhood. As Rosemarie herself had observed, this kind of thing just didn't happen in Hillsdale. She spoke frequently to the media, hoping that someone who might have seen something would come forward. Her main memory of the media session was the dirty footprints that had turned the light-brown carpet on the steps a charcoal gray.

The anxiety in the D'Alessandro house that night was almost unbearable. Frank often displayed anger when he was frustrated. The night before, he had an outburst over not having a box in which to wrap an Easter gift. "He could be calm and patient for long periods and then change in a moment," Rosemarie recalled. "He had a good job, but he wasn't communicative, and he was never really my soul mate."

Hillsdale police chief Philip Varisco was on vacation in Florida when he was informed of Joan's disappearance. Hillsdale was the kind of community, and Varisco was the kind of leader, that made it unthinkable that the chief not be present for a trauma like this. He rushed back home. Varisco, who passed away in 2012 at the age of eighty-nine, was a complete professional. He attended the FBI National Academy program in Quantico to make himself and his force as effective as possible.

The chief went to the D'Alessandros' house the next day. Rosemarie was sitting on her front steps when he came up the walk. He told her he was taking personal charge of the investigation. While not promising the happy outcome he knew was unlikely, he calmly assured her that everything would be done the right way. He asked for a photograph he could release to the newspapers. Rosemarie went to a picture of Joan in her school uniform, hanging in the hallway, took it off the wall and out of its frame, and handed it to Varisco.

Frank told newspaper reporters that if whoever had taken Joan would bring her back safely, he would ask the authorities to waive prosecution. During a television interview, Rosemarie described Joan to reporter Vic Miles, how special she was and how much she was loved, pleading for her to be returned. Years later, one of Joan's classmates told Rosemarie she remembered the broadcast as if it were yesterday because Joan's mommy was asking on TV for her to come back. Only two months before, Rosemarie had suddenly had the terrible thought of what would happen if one of her children died, and how utterly, unimaginably heart-wrenching it would be.

The police questioned several possible suspects, including a man seen driving around the neighborhood about an hour before Joan disappeared and another wandering the area on foot. The first turned out to be looking at neighborhoods in which to move and the second was simply lost. There are almost always loose ends and red herrings in major cases. But the investigators focused quickly on Joseph McGowan. Though he had no criminal record, it had been his house to which Joan said she was going and Rosemarie had related her creepy encounter with him. Her father had seen him taking out the trash the day after Joan's disappearance and, pointing to the house on the corner, said to Rosemarie, "Something's not right over there."

Police officers and detectives spoke to McGowan both Friday and Saturday, asking him to account for himself in the minutes and hours after Joan went to his house. He was calm and amiable but denied that he saw Joan on Thursday. Instead, he claimed he was at the nearby supermarket buying groceries at the time Rosemarie said her daughter went to his house. What about the car Joan saw pull into the driveway? Did anyone see it leave the garage? No, he walked. At which cash register did he check out? He didn't remember. Could he show them the grocery receipt? He thought he threw it away. Might it still be in the trash? He thought the trash had already been picked up. What days did the trash collectors come? He wasn't sure. What did he buy? Steaks and apples, among other things. Were the steaks still in the refrigerator? No, he and his mother had eaten them. What about the apples? He wasn't sure.

Seasoned detectives develop a natural sense for knowing whether a suspect's story and profession of innocence is true. Over lunch one day, Mark Olshaker asked retired LAPD detective Tom Lange when he came to his own conclusion that O. J. Simpson was the prime suspect in the murders of his ex-wife Nicole and her waiter friend Ronald Goldman in 1994. Lange replied that though O.J. was cordial and cooperative during the interview, he asked no questions regarding the details of Nicole's death, whether or how much she had suffered, whether the police had any idea who had done it—all of the normal things any close survivor would instinctively want to know.

Joan's friend Rich recalled a large crowd in front of the police station on Central Avenue as McGowan was inside being questioned. To his young eyes, it seemed as if the entire town had gathered there.

As the seeming holes and contradictions in McGowan's statements grew increasingly glaring, the detectives asked him to submit to a polygraph examination at the station. He agreed.

McGowan failed the polygraph, and when detectives informed him, they confronted him with all his statements that didn't add up. Finally, exhausted and without any more answers, he asked for a priest. He and the priest met privately, and McGowan confessed to him. He then confessed to the detectives, and told them that after he killed Joan, he drove her body across the New York state line and deposited it in Harriman State Park in Rockland County, about twenty miles away.

Chief Varisco took it upon himself to be the one to tell Rosemarie and Frank. It was a little after four P.M. A deeply sensitive man, he brought a Catholic priest with him, and together they sat with Rosemarie at the kitchen table. Rosemarie remembers removing the tablecloth from the white table to delay, if only for a few moments, what she knew was to come.

When the chief told her what McGowan had said, she cried, "I want to kill him!" She says she was feeling rational and in control as she said it, knowing at the time she didn't mean it, but needing a way to release the anguish consuming her.

The priest admonished her not to speak that way.

"What do you expect, Father?" Varisco said.

2

"I SLEPT WELL"

Dr. Frederick T. Zugibe, the chief medical examiner for Rockland County, New York, said Joan's case was one of the most emotionally difficult in his long and distinguished career.

Word had spread quickly from the Hillsdale Police Department to the Bergen County district attorney's office, and from there to the Police Division of the Rockland County sheriff's office. So early in the afternoon on Easter Sunday, Officer John Forbes drove to the location described to him in Harriman State Park, just off Gate Hill Road, near the southern end of the park.

There he found the naked and battered body of a young white female. She was faceup in a wedge-shaped crevice between two boulders, on a leafy slope under a rock ledge. Her head was twisted sharply to the left and she was facing down the slope. Forbes had four young children of his own and struggled not to break down.

He called in the crime scene team.

By the time Dr. Zugibe arrived, less than an hour later, the cordoned-off crime scene was already populated by a horde of police officers and crime scene technicians, detectives, FBI agents, reporters, press photographers, and the generally curious. He immediately ordered the officers to move all nonessential personnel away.

Richard Collier, the D'Alessandro neighbor who was an FBI special agent working out of the New York City field office, came in to identify the body.

Yes, it was Joan.

Though the crime scene was no longer pristine, her body had not been moved or touched. Dr. Zugibe immediately noted the lividity—the purpling of the flesh around her abdominal area. This told him she had not been killed at this spot. If she had been, the lividity would have been concentrated in her back due to gravity. Since that type of blood settling takes at least six hours, he also knew that she hadn't just been dumped here. He took her body temperature and found that it matched the temperature of the air. That indicated she had been dead at least thirty-six hours, the time it takes for a body to cool completely. The finding was confirmed for him by the absence of rigor mortis, a postmortem stiffening of the muscles that begins several hours after death and subsides within twenty-four to thirty-six hours.

Taking all the observable physical evidence together, Dr. Zugibe estimated that Joan had been dead for about fifty hours. When he was able to do more sophisticated tests during the autopsy, he upped his estimate to a minimum of seventy hours, which meant that she had died within a couple of hours or less of when Rosemarie last saw her.

Sheriff's officers conducted a thorough search of the surrounding area and found a gray plastic shopping bag printed with the Mobil logo. According to Zugibe, the bag was neatly packed rather than haphazardly stuffed and contained the clothing Joan had been wearing when she disappeared: a pair of red and white sneakers, a turquoise shirt, maroon pants, white socks, and white panties, stained red with her blood.

Before the body was removed, an officer called the Marian Shrine in Stony Point, New York, and asked that a priest come to the scene. When he arrived, illuminated by the police lights and in the presence of officers, detectives, FBI agents, and reporters, he administered last rites to Joan Angela D'Alessandro. Once the priest was finished,

Zugibe officially pronounced death, a seemingly obvious observation but a necessary formality in any murder investigation.

Back at the medical examiner's office in Pomona, New York, less than ten miles away, he began the autopsy. From my experience in dealing with many medical examiners over the years, I would say there is little that is more painful than having to examine a dead child, and absolutely nothing more agonizing than if the child has been murdered.

By the time he had completed the postmortem, Zugibe listed injuries that spoke to the utter depravity of the crime: fracture of the neck, manual strangulation, dislocated right shoulder, generalized deep bruising, lacerations under the chin and inside the upper lip, frontal fracture of the skull, fracture of both sinuses, swelling of the face, both eyes blackened and swollen shut, three teeth loosened, contusion and hemorrhaging of the brain, bruising of the lungs and liver, and rupture of the hymen.

Essentially, Joan was beaten, choked, sexually assaulted, and ultimately battered to death. But according to Dr. Zugibe, it was even worse than that. Had she died right after the beating and strangulation, her face and body would not have appeared swollen. Upon death, the homeostatic functions that cause swelling at an injured site shut down. And since swelling takes about half an hour to be completed, he concluded that Joan must have been alive for at least that long following the attack. Mercifully, she was almost certainly unconscious.

The medical examiner's close examination of the neck revealed two areas of injury: the thyroid cartilage and the hyoid bone. His conclusion was that a half hour or so after the deadly attack, the offender, unsure that he had killed her, returned to finish the job with a second manual strangulation. This sounds completely believable to me. With someone like Joseph McGowan, an "inexperienced killer," it would not be unusual for him to be unsure how effective he had been in dispatching his victim and wish to take no chances.

I had seen a similar sort of behavior in the Christmas 1996 murder of six-year-old JonBenet Ramsey in her home in Boulder, Colorado. The medical examiner's report listed two potentially lethal injuries: blunt force trauma to the head and ligature strangulation. Since there was no bleeding at the crime scene, I concluded that the cause of death was the strangulation and that the severe blow to the head was an attempt to make sure she was dead.

This scientific evidence suggested something highly significant from a behavioral perspective. No parent without a history of extreme child abuse could possibly, and systematically, strangle that child to death over a period of several minutes. It just doesn't happen. Taken together with all of the other forensic and behavioral evidence, this did not tell us who killed JonBenet. But it told us who *did not* kill her: either of her parents. Mark and I came up against a lot of pushback and public condemnation for this conclusion, including from my old FBI unit, but the pursuit of criminal justice is not a popularity contest, and you have to let the evidence speak for itself.

Which was precisely what I would do with Joseph McGowan.

JOSEPH MCGOWAN WAS ARRAIGNED BEFORE BERGEN COUNTY JUDGE JAMES F. MADden. He was held in lieu of the $50,000 bail set by the judge. On Tuesday, April 24, 1973, he was indicted for the murder of Joan D'Alessandro.

Two days later, in the late morning, Joan's funeral was held at St. John the Baptist Roman Catholic Church, whose school Joan attended. The children from her class were there, and after the service, they all lined up outside as her casket was carried out to say goodbye.

As an investigator of violent crime, you try to be as emotionally detached as possible, not only to maintain your objectivity and critical judgment, but also to preserve your sanity. In fact, having to put myself, as a behavioral profiler, into the head of every victim whose case I work has definitely taken its psychic toll on me throughout my career. Dr. Zugibe's and Officer Forbes's reactions to seeing Joan's small body in the park were understandable. No matter how "professional" you try to be, you can't *not* react to something like this.

What kind of a man or monster spontaneously does something like this to a seven-year-old girl? I asked myself as I read through the case file a quarter of a century later. That's what I would seek to find out.

McGowan repeated his confession to Dr. Noel C. Galen, a forensic psychiatrist who had received his neurology and psychiatry training at Bellevue Hospital in New York and consulted for the New Jersey court system. The day after his indictment, McGowan detailed for Dr. Galen how he answered the door, and when Joan told him what she was there for, said she should come with him downstairs so he could get the money for her. She must have hesitated or resisted, because he admitted grabbing her and forcing her into his downstairs bedroom. Meanwhile, McGowan's eighty-seven-year-old grandmother, hard of hearing, was watching television upstairs. His mother was at work.

I am not revealing any privileged information from McGowan's case file or medical records. All of the evaluations and analyses I am citing were contained and published in the appellate decision *Joseph McGowan, Defendant-Appellant, v. New Jersey State Parole Board, Respondent*, decided by the Superior Court of New Jersey on February 15, 2002.

As he told Dr. Galen, once in the bedroom and "safely" away from the street, McGowan ordered Joan to take off her clothes. Though he said he "never completed the act," he became sexually excited and ejaculated on his fingers only inches from Joan and then penetrated her digitally. He probably couldn't wait and did this before she was completely undressed, as her panties were stained with blood. Since he acknowledged he had semen on his fingers, we can't say for sure whether he "completed the act" or not, but the blood and injuries to her vaginal area indicated a brutal assault.

It was at this point, according to McGowan's account, that the consequences of his impulsive action dawned on him. "All of a sudden," he told Dr. Galen, "I realized what I had done. If I let her go, my whole life was gone. All I could think of was just to get rid of her."

As an investigator, I have to say that from a criminological perspective, this part makes sense to me. In a high-stress situation like this, a "smart" offender will tend to have only one thing on his mind: getting away with his crime. This is apparently what happened with McGowan. Whether Joan lived as long after the first strangulation as Dr. Zugibe speculated is an open question, as is which of McGowan's attempts to kill her was the successful one. But the general narrative of what happened is not in doubt. According to the transcript of the confession:

> *I grabbed her and started to strangle her and I dragged her off the bed, tossed her into the corner of my room on the tile floor, off the rugs. She was trying to, you know, scream, and was fighting back. But of course she really couldn't, since I had my hands around her throat. Uh . . . she stopped struggling . . . just sort of lay there. I got dressed. I had been sweating so violently. I went out to the garage. I got some plastic bags to put her in. [Returning from the garage,] I saw that she was still moving, so I began strangling her again and I hit her head on the floor repeatedly. She began to bleed from the nose, mouth, face . . . I don't know where. There was blood all over the floor. I then grabbed one of the plastic bags and put it over her head and twisted it tightly and held it there until she stopped.*

AS I READ THIS IN PREPARATION FOR MEETING HIM, I WAS THINKING TO MYSELF: *AN* *hour or two earlier, this guy was standing in front of a classroom teaching chemistry to high school students. What led from that Point A to this Point B?*

As the confession went on, McGowan described how he lifted Joan's body and placed it in a plastic trash bag, then wrapped it in an old sofa cover, tied it with cord, carried it to the garage, and stowed it in the trunk of his car—the "new car" Joan had spotted from her front yard around the corner and down the block. He cleaned up her blood as best he could with some old T-shirts. Then he drove the twenty miles or so and dumped the body down the slope in Harriman State Park. He unwrapped it and left it under the rock ledge. He put the plastic bag and the couch cover in a roadside trash can.

When he returned to Hillsdale he joined the neighborhood search for Joan.

“I felt better when I went back to the house,” he told Dr. Galen. “I slept well.”

3

MIND OF THE KILLER

Frank Mikulski, who retired as Hillsdale chief of police in 2006 after forty-two years on the force, was a patrol sergeant when Joan was killed.

"It was the most horrible crime that ever occurred in the borough and it's the one that still stands out in my mind," he reflected to the Bergen *Record*. "This man was a monster, and when something like that happens to a child it's embedded in the memory of the community and it never goes away. For people here, it's like Pearl Harbor or 9/11 . . . You remember where you were and what you were doing."

Just about everyone who knew either the D'Alessandros or Joseph McGowan remembers where they were and what they were doing when they heard the news.

Robert Carrillo, a math teacher who rode to work with McGowan and one other teacher, had thought of McGowan when Joan went missing. "When the news came over, the first thing I thought of was Joe. I thought, *Gee, he lives there. I wonder if he knows her, not that he would have been involved.*"

On Easter Sunday, Carrillo had gone to visit his mother in Queens with his wife and daughter. That evening they were returning home when they heard. "We were on the Cross Bronx Expressway and they made an announcement on the radio that they had caught the suspected killer of Joan D'Alessandro, and he was a high school science teacher from Rockland County and they gave his name. And I had to pull off the side of the road. I got physically ill."

Jack Meschino taught chemistry with McGowan. He and his longtime partner Paul Coletti had socialized with him and other teachers on several occasions. "I remember when we heard," Coletti recalled, "when we received the phone call, hanging up, and we just sat and looked at each other and said, '*What!*' "

"Yeah, it was just unreal," Meschino agreed. "It was really a shock that he did something like this." But then he added, "On the other hand, Joe was a strange fellow; he really was. And in retrospect, you think of things. Another thing that struck me was Joe's humor. There was a big gap in it. Things that he laughed at and thought were funny, the general population wouldn't laugh or go along with. Very bizarre.

"Joe always walked around with a set of keys—more keys than anyone would have need of. God only knows what they were for. One of the things Joe took it upon himself to do was to check classroom doors after the school day ended. And it was known that he actually turned in some of his colleagues for leaving their classroom doors unlocked. This was not in his line of duties. He had no administrative responsibilities. The only people he tried to ingratiate himself with were the administrators."

"He was kind of seen as an administration toady," said Carrillo. "Also, I remember when the in thing to do was to join the Playboy Club in New Jersey, and Joe was a gold card member. He made it a point to show everyone in the teachers' room his gold card. These kinds of things were important to him, seeking out approval or recognition from others."

We asked Carrillo if McGowan was popular with the students. "I think he was," he responded. "He was the type of teacher who tried to be friendly with them. He strove to be liked by the kids."

It didn't always work out that way, though. Mark and I later learned from a number of his female students that they had felt uncomfortable around him. One woman, now in her early sixties, recalled in a chemistry lab asking Mr. McGowan what to do with a glass flask she no longer needed. McGowan grabbed it from her and threw it to the floor, where it shattered across the room. He offered no explanation for his action.

Other students had similar thoughts about McGowan. One shared her story on social media: "Back in the day, I believe senior year 1971, I had McGowan for chemistry. I was so creeped out by him that I went to the office and demanded to get out of his class."

Tappan Zee High School was closed for spring break the week after Joan's murder, but when classes resumed the following Monday, an atmosphere of stunned silence prevailed.

"Going back to school was very bizarre," Carrillo said. "Everybody knew about it, but nobody really talked about it much. The students may have talked about it among themselves, but the staff was more in shock than anything else. The board of education dismissed [McGowan] at a very private meeting; there was no publicity about it."

Carrillo and Eugene Baglieri, the other teacher in the car pool, talked about it, "and you think of things, looking back," Carrillo said. "But it was such a terrible experience, even to be touching on it in a peripheral way, that people just avoided it."

For Jack Meschino, it was even worse. "It was terrible coming back," he remembered. "We team-taught, and all of a sudden, *our* students were *my* students. I'll never forget the first few classes. It took me five or ten minutes even to get up the courage to address the students. We all sat there and looked at each other; didn't know how to handle it. We were dumbstruck."

There would be more psychological exams for Joseph McGowan in the next several weeks as he sat in the Bergen county jail in Hackensack. On May 10, 1973, a little more than two weeks after Dr. Galen's interview, Dr. Emanuel Fisher, a psychologist, tested the suspect. He found McGowan to be "an exceedingly labile, tense and hysterical personality whose tendency is to act out mood and impulse in a very explosive manner. Rational controls are weak, despite the fact that he is an exceedingly brilliant individual."

Dr. Fisher noted "a tremendous amount of underlying, unconscious hostility," that he "repressed, avoided, sublimated and intellectualized." And although he presented himself as "a very proper, conventional, conforming individual, [t]his exaggerated propriety, conventionality and conformity, constitute his defensive facade, for himself and for others, against the underlying depression and hostility of which he is unconscious."

Less than a month later, on June 6, Dr. Galen submitted a psychiatric report based on his interviews with McGowan. He said McGowan had "given a well-documented history of sexual attraction to young girls. This, coupled with a clearly evident picture of a dominating and overprotective mother, would strongly hint at some profound problems in [relation] to his making a normal adjustment to an adult woman."

Galen cited McGowan's admission to him that for the previous year or so, he had found himself sexually aroused by young girls and specifically mentioned his twelve-year-old female cousin. He said he had masturbated to rape fantasies. From this, the psychiatrist concluded, "younger girls would pose no threat to his rather shaky concept of his manhood."

A further neuropsychiatric report was submitted by Dr. Abraham Effron in October, confirming what McGowan had told Dr. Galen about "fantasies of sexual relations with little girls," adding that he was sexually aroused as a nineteen-year-old camp counselor when a young girl sat on his lap.

Dr. Effron also interviewed McGowan's mother, Genevieve, who had not been home at the time of the murder. She said that her husband, who had died of a heart attack while Joseph was in college, "was much closer to the boy [than she was] and would take him out often."

After Joe completed college, he moved back home with her and his grandmother.

Effron's report stated:

> *He does not show what he necessarily feels. He conceals many facets of his complex true self and his true identification and related emotional difficulties. He tries to hide his inability to truly establish his masculinity. He experiences tension whenever he gets close to the opposite sex. This passivity generates anxiety, which in turn feeds on itself and results in a higher state of tension, which must expiate itself in a complete loss of self-control or sexual release.*
>
> *He manages to control an underlying psychosis by intellectually holding in abeyance his primitive drives to an inordinate extent, but as in the past and tragically in the recent past, he may act out again.*

There is an ongoing debate regarding whether violent predators are born or made—the so-called nature versus nurture question. I would argue that, while no one who does not have certain inborn tendencies toward impulsivity, anger, and/or sadistic perversions is going to evolve into a predator because of a bad upbringing, there is no doubt in my mind that those possessing such inborn tendencies can be pushed along the path to predation by negative influences as they grow up and mature.

In fact, Ed Kemper was one such individual.

EDMUND EMIL KEMPER III WAS THE FIRST PRISON INTERVIEW THAT BOB AND I DID after my idea to speak with the killers. The only problem was we didn't really know what we were doing.

As FBI agents, we were pretty much trained on the job to interview witnesses and interrogate suspects. But neither of those skill sets really prepared us for the prison interviews. An investigative interview is a meeting with one or more persons who may have information relative to a crime or the perpetrator of that crime. We try to find out as much of the who, what, when, where, why, and how as possible. That person is not treated as a suspect.

An interrogation, on the other hand, involves questioning a potential suspect in a crime. That individual is entitled to be informed of his or her legal rights and in no case may the information violate rules of due process. This tends to be more of a show-and-tell on the part of the interrogator in which the suspect is advised of or shown definitive forensic evidence linking him to the crime. The questions fall into the form of not if, but why and how, to get the suspect to cooperate and confess.

Neither of these approaches was appropriate for our prison interviews. The interplay between the agent and the violent offender needed to be informal and not overtly structured. What we were looking for was not so much the facts of the case, which were already established, but the motivation, the pre- and post-offense behavior, the victim selection process, and then the big question of why, without being too assertive, directed, or leading—the opposite of what we'd try to do in a suspect interrogation.

As counterintuitive as it sounds, the prison encounter had to feel "natural"—just a couple of people talking freely and exchanging information.

Since we were in California, we decided to go after the local "clientele" first. A special agent out there who was one of Bob's former students agreed to act as liaison for us with the state prison system. Ed Kemper was a six-foot-nine, 300-pound giant of a man who was serving multiple life sentences at the California Medical Facility at Vacaville, midway between Sacramento and San Francisco. Kemper had become known as the Coed Killer for his string of murders in and around the University of California, Santa Cruz in 1972–1973.

Before the interview, we familiarized ourselves with all the details of his grisly record. This would become a standard part of our process, so we wouldn't be misled or conned by men who made a specialty of the practice. What we wanted was not the facts so much as what guys like Kemper were thinking and feeling as they planned and executed their crimes. We wanted to know what motivated them, what techniques they used, and how they regarded each assault or murder afterward. We wanted to know how and where the fantasy began, what the most emotionally satisfying parts of the crime were, and whether torture and the suffering of the victim were important components for them. In other words: What were the distinctions between the "practical" aspects of successfully committing the crime and the "emotional" reasons for doing it.

Born in 1948, Kemper grew up in a dysfunctional family in Burbank, California, with two younger sisters and parents Ed and Clarnell, who fought constantly and eventually separated. Early on, Ed showed a predilection for dismembering the family cats and playing death ritual games with his sister Susan. Clarnell sent him off to live with his father, and when he ran away, she sent him to live with his paternal grandparents on a remote farm in the foothills of the California Sierras.

Told by his grandmother Maude one day to stay and help her with the household chores rather than accompany his grandfather Ed into the field, the hulking fourteen-year-old shot Maude with a .22-caliber rifle and then stabbed her repeatedly with a kitchen knife. When Grandpa Ed returned home, the boy shot him, too. This got him committed to Atascadero State Hospital for the criminally insane, until at age twenty-one and over the objection of state psychiatrists, he was placed in Clarnell's custody.

Sitting calmly in the prison interview room, Kemper took us through his childhood and his mother's fear he would molest his sisters, so she made him sleep in a windowless basement room, which terrified him and made him resentful of his mom and sisters. That was when he mutilated the cats. He described his succession of odd jobs once he got out of Atascadero, how Clarnell, a secretary at the newly opened University of California, Santa Cruz, was popular and caring with students, but gave him the message that he would never be in the league with the beautiful coeds who attended there. He described his habit of picking up beautiful female hitchhikers of the type he'd missed out on by being imprisoned throughout his formative years, and how this habit eventually evolved into abduction and murder. He told us how he would bring the bodies back to his mother's house, have sex with them, then dismember them and dispose of the pieces. Though his victims certainly suffered horribly, he was not motivated by sadism as many serial killers are. What he told us he was doing—and this isn't a phrase I've heard before or since—was "evicting them from their bodies" so he could possess them, at least temporarily, after death.

And then he related how, after two years of this, on Easter weekend he'd finally summoned the will and courage to go into his mother's bedroom while she slept and bludgeon her to death with a claw hammer. He then decapitated her, raped her headless corpse, cut out her larynx, and fed it down the garbage disposal. But when he turned on the switch, the device jammed and threw the bloody voice box back out at him. He took this as a sign that his mother was never going to stop yelling at him.

He called a friend of his mother and invited her over to the house for dinner. When she arrived, he clubbed and strangled her and cut off her head. He left her body in his bed while he slept in his mother's. On Easter Sunday morning, he took off and drove aimlessly until he reached the outskirts of Pueblo, Colorado. He stopped at a phone booth, called the Santa Cruz police department, took some pains to convince them that he was the Coed Killer, and waited to be picked up.

Kemper was lonely and narcissistic and wanted to talk to the point that at times I had to tell him to stop because we had specific questions to ask him. We used a handheld tape recorder and took notes. This was a mistake. We learned that because we had taped the interview the subject lost a measure of trust in us. These guys are mostly paranoid by nature, but in prison, there are good reasons for that. There was worry that we would share the recording with prison authorities or it would get out to the general population that a prisoner was talking to the feds. The notes were not a good idea, either, for much of the same reason. And the subject expected us to give him our full attention.

Still, despite these necessary adjustments, much of that first conversation gave us significant insights. Perhaps most important, it demonstrated from the start just how pertinent this question of nature versus nurture would be when it came to understanding what drove these men in their antisocial behavior. This issue would come to infuse just about every interview that I've ever done with a killer, and the same would likely be true with Joseph McGowan.

While McGowan did not suffer the same emotional trauma growing up as Ed Kemper had, his domineering and controlling mother clearly had a profound effect on his development. He was a highly intelligent twenty-seven-year-old teacher with a master's degree in science, yet he was living in his mother's basement and was still emotionally dependent on her. His inability to go against her and then being forced to live with her as a mature adult surely had an impact on his self-image and, as I would discover, the life of an innocent little girl.

IN A BERGEN COUNTY COURTROOM, WITH A JURY ALREADY IMPANELED, MCGOWAN AND his attorneys decided to forgo a trial and instead entered a guilty plea to first degree felony murder on June 19, 1974. From his perspective, I think that was probably a wise decision. Given the facts of the case and the certainty of his guilt, I can't imagine a jury regarding him with any compassion or leniency when it came time for sentencing.

On November 4, New Jersey superior court judge Morris Malech sentenced him to life in prison, with the possibility of parole after fourteen years. McGowan had his lawyer try to appeal this sentence multiple times, but all of his attempts failed.

The following month, McGowan was examined by another psychiatrist, Dr. Eugene Revitch, at the New Jersey Adult Diagnostic and Treatment Center in Avenel. Dr. Revitch, trained in both psychiatry and neurology, was a clinical professor at the Robert Wood Johnson Medical School of Rutgers University and published some of the first papers on sexual assault and murder.

Once again, McGowan admitted to rape fantasies in college, caused by sexual frustration and anxiety. After listening to the account and examining his subject with and without sodium amytal (so-called truth serum) and finding little difference other than degree of affect, the psychiatrist stated that Joan's homicide was "not a cold-blooded murder, but something committed in a state of extreme emotional disorganization and pressure. The killing was the consequence of an additional upset and failure due to premature ejaculation." Dr. Revitch also recognized "a degree of dissociation with use of mechanism of denial."

While I have seen some cases of rape turn into murder as a result of premature ejaculation or failure to achieve or maintain erection on the part of the attacker, it tends to be with two specific rapist typologies—the anger-retaliatory rapist and the exploitative rapist. These guys tend to focus on adult women as their victims, and if the premature ejaculation or similar embarrassment results in either a mocking response from the victim or a loss of face for the attacker, then the situation can turn dire. Given that the victim here was a child, I was pretty convinced that was not what we were seeing here. But it was Dr. Revitch's conclusion that really had me wondering:

> *We believe these events only occur once in a lifetime of such individuals. A series of circumstances are necessary to provoke the incident. If the girl had not come to his home that day or, perhaps, if he had two dollars instead of only one*

dollar and a twenty-dollar bill, the event would not have taken place, at least at the present.

Clearly, the crime would not have taken place had Joan not showed up at the McGowan house and rung the doorbell. She was a tragic victim of opportunity. Beyond that, from what I knew from my own study of the criminal mind, I wasn't sure how much I agreed with the various psychological reports.

Which evaluation was closer to the mark: Dr. Effron's opinion that "he may act out again," or Dr. Revitch's conclusion that "these events only occur once in a lifetime of such individuals"?

I reserved judgment until I actually talked to McGowan myself.

4

HUMAN FALLOUT

If there is one word I've found that survivors of murder victims detest, it is *closure.* The media, the public, well-meaning friends, and even the judicial system itself often feel that this is what all grieving loved ones are seeking so they can "put this behind them and get on with their lives."

But anyone who has "experienced" a murder knows there is no such thing as closure, nor in fact should there be. The mourning process will go through stages and eventually the pain will become less unbearably acute, but it will never go away, any more than the hole in one's personal universe left by the loss of the victim and the erasure of a lifetime of promise will ever be filled in.

The Girl Scouts sent a sympathy card. Other than that, no one in any official capacity made the effort to contact the family.

The emptiness really started, Rosemarie says, "after the burial, when everyone left and went back to their own lives." The most important thing for her at that point was to keep life as normal as possible for Frankie and Marie. "I made sure Marie stayed in Girl Scouts because she wanted to, even though it was painful for me even to think about Girl Scout cookies. We stayed in the same house, so the familiar things would still be there, like school and friends, and not having more changes to deal with. I tried not to be overprotective. I let them continue to go out and play, though I was always attentive to their whereabouts. They had to be children and I didn't want to be paranoid."

Nor would she shield them from the ongoing news about their little sister's case. "I would tell them both what was going on so they would hear it from me. I knew they would be hearing things and I didn't want them to hear in a scary way. We would sit on the bedroom floor and talk about anything they had on their minds. They looked forward to that and knew they weren't being left out." Rosemarie and Frank took them to the cemetery on various occasions to visit their "sister in heaven."

Rosemarie came to accept the reality that her love and pain could not be separated. "I have felt a relationship with Joan in my heart all these years," she says. "It's a relationship I wouldn't have chosen, but it's still a relationship that inspires me to do what I do. And I found there is a peace that comes with that."

It was not all Rosemarie had to contend with. Seven months after the murder, her beloved father died of cancer. He had adored his granddaughter and never stopped grieving for her.

Rosemarie went to the court hearing when McGowan pled guilty. She felt she had to be there for Joan. Genevieve McGowan was there, too. "When I walked into the courtroom, she gave me the coldest stare I have ever experienced in my life. It was the first time I had ever seen her."

Rosemarie would not have minded if McGowan had stood trial so that the truth would have come out and nothing would have been held back, including the details of what had happened to Joan. But as often happens, other types of details began filtering back to her. One of the most appalling was when she heard through a friend that Genevieve had told an acquaintance from church that she hated Rosemarie, because if it hadn't been for her, Joe wouldn't have killed Joan and gone to prison.

And then there were the persistent challenges from her own body. The first inkling, when she stopped to think about it, had come years earlier, when she was a nineteen-year-old in New York. One day she was running for a bus. Suddenly her leg felt tense and then gave out on her. She didn't know what it was, but it didn't recur, and she didn't think much about the incident.

A few years later, when she was pregnant with Marie, she was feeling exceptionally tired and knew it wasn't the normal fatigue of pregnancy.

She tried to work out her own strategies and coping mechanisms for dealing with her unknown affliction. "I had to develop my own strength through focus and determination."

When Joan was born, the tiredness became more pronounced and she was forced to hire help, which she kept until the baby was six weeks old. The ongoing symptoms were vague and fluctuating, and seemed to affect various regions of her body. "It got worse as the day went on and was worst in the late afternoon. I knew I had *something*." The one common denominator was the extreme fatigue and the knowledge that she had a limited amount of energy on any given day, and if she used it up, there would be consequences going forward.

She went to doctors, but they couldn't find anything. Or they told her it was a physical manifestation of postpartum depression. Or it was a virus and she would get over it. But she didn't get over it, and "if I didn't rest, I would pick up infections regularly."

It wasn't until a year after Joan's death that Rosemarie finally got an accurate diagnosis. She checked into Mount Sinai Hospital in New York and submitted to an extensive battery of tests. A neurologist there concluded that she was suffering from myasthenia gravis, a neuromuscular disease caused by the breakdown in normal communication between nerves and muscles. It is an autoimmune disorder that can be related to thymus gland abnormalities, with minimal or no relationship to one's genetic background. There was and is no cure, and treatment centers on trying to alleviate symptoms that, in addition to the severe fatigue and weakness, can involve drooping eyelids, double vision, slurred speech, difficulty chewing and swallowing, and even trouble breathing.

"They told me every case of MG is different. If I pace myself and stay organized, it's a little better," she says, but adds, "I do take risks, and that is when I get the most joy out of life. In fact, I think it makes me appreciate moments of joy so much more because this condition makes the experiences so pointed."

After a miscarriage, two of those joys occurred in 1980 and 1982, when Michael and John were born. Frankie and Marie were already in their teens, so for Rosemarie and Frank, it was like having a second generation of children.

But the joy would not last. Frank lost his job, and their marriage started to fail. "Even though he found another good job, he was lashing out at us more," Rosemarie says. "And when John was eight, I witnessed inappropriate touching and other gestures." Throughout all the trials, Rosemarie was sustained by her religious beliefs and devotion. "In my faith," she commented, "God was always my psychiatrist. After what happened to Joan, I asked Him to help me choose a life without animosity, and instead, a life advocating for prevention, protection and justice."

In the early 1990s, around the time Michael was eleven and John was nine, Frank moved downstairs. Rosemarie knew it was just the next step on a road that led in only one direction. "I was going to go for a divorce in 1993—on September 7, Joan's birthday," she related.

Then they received a telephone call that changed her life yet again. It came on July 26 of that year, from Deputy Chief of Detectives Ed Denning from the Bergen County prosecutor's office. He said that Joseph McGowan was coming up for parole. This was a shock, because Rosemarie had not been informed that six years could be cut for good behavior and work credit. He had been turned down in 1987, the first time he'd been eligible, but his chances looked better this time since he'd served more than his minimum sentence.

It had been twenty years since the murder, and Rosemarie wanted to bring Joan back into the public consciousness. "It wasn't about dwelling on the grief but being a squeaky wheel to fight to keep her killer in prison and trying to make sure he wouldn't be up for parole every few years. I thought starting a movement of the people would help us all."

The mother of a former Tappan Zee High cheerleader called Rosemarie to say her daughter felt she had been stalked by McGowan back in school and would be "petrified if he came out."

Rosemarie knew she'd have to fight to keep him behind bars and began by working with local and county officials, district representatives and the community to organize a vigil on September 30, 1993, at Veterans Park in Hillsdale. More than a thousand supporters attended. "My divorce plans had to change according to the advice that I sought from an attorney," she explains. "The focus had to be on the fight to keep McGowan in, and that couldn't be complicated by the divorce."

And she had two overwhelming reasons to keep him behind bars: to make the punishment at least in some sense commensurate with the enormity of the crime; and to make sure no other young child suffered at McGowan's hands as Joan had. If there were to be any meaning to Joan's death, any meaning to her disappearing on Holy Thursday and being found on Easter Sunday, Rosemarie understood, she would have to do something herself. "The message of hope was clear. It would be the movement for child protection and helping society inspired by Joan, Holy Thursday, and Good Friday." It was as if God, who had ordained free will to mankind and therefore had to suffer the deaths of little children at the hands of those who would forsake His values, was giving her a message.

"I realized then that this is the work I'm supposed to do. And I saw it as getting closer to Joan's spirit. That's when the movement started. I didn't get positive support from my family—instead quite the opposite when family members verbally assaulted me, actually threatened me with physical force, and sent harassing mail. In the late 1990s, Michael and John would become involved, but before that, I was pretty much on my own."

She began speaking out. She began organizing. She spearheaded a nine-month campaign to make the public aware of the danger of child predators and the reasons that they should be kept in prison. The parole board listened and once again turned down McGowan's request. Just as important, it reviewed his case and had him transferred to the maximum security facility in Trenton, where the warden felt he should have been to begin with. Additionally, the board imposed a future eligibility term (FET) of twenty years before his next hearing. With good behavior and work credits, this would be reduced to twelve years, making him once again eligible in 2005.

And Rosemarie did not just let matters stand once McGowan's 1993 parole request was denied. She began a grassroots movement, organizing parents and other interested parties in rallying and petitioning for justice for child victims. She wrote; she called; she appeared on television and radio and sat for interviews. Wherever she went, she handed out little green bows, Joan's favorite color.

It took three years of essentially full-time advocacy. Then, on April 3, 1997, Governor Christine Todd Whitman signed what became known as Joan's Law. Wearing a green bow on her lapel in memory of Joan, Governor Whitman sat in bright sunshine outside the Bergen County jail. Rosemarie, Frank, and Michael and John stood around her, surrounded by police officers, detectives, sheriff's deputies, and legislators, all of whom had supported the campaign to have the law enacted.

Joan's Law amended the New Jersey criminal code to mandate that anyone convicted of the murder of a child under fourteen years of age during the commission of a sexual assault would be sentenced to life in prison without possibility of parole.

Rosemarie came to the podium and thanked the governor and the sponsors and supporters of the bill. "Maybe this can deter crime—we hope," she stated. Then she held up a photograph of Joan and said, "It's she who we have to be thankful for. Joan's spirit is very much alive. She wants you to smile more. She wants you to be more positive."

The following year, on October 30, 1998, President Bill Clinton would sign a federal version of Joan's Law. Six years after that, on September 15, 2004, New York governor George Pataki traveled to Harriman State Park, the site where Joan's body was found, to sign a Joan's Law for his state. Rosemarie could not make the signing but listened on the phone from her bed. It was where she had made many of her calls to connect with people to get bills passed.

Ironically, one convict Joan's Law would not affect was Joan's killer, Joseph McGowan. He had been sentenced before the statute went into the code, and the law could not be made retroactive. So, according to the instructions of the court of appeals, the New Jersey parole board and Rosemarie D'Alessandro prepared for the next hearing.

This was particularly problematic because shortly after the 1993 decision was handed down, McGowan appealed the ruling that didn't allow him a parole hearing until 2005. The appellate court requested additional information from the parole board, then let its ruling stand. Over the next few years McGowan appealed three times, and the D'Alessandros went to each one. Their victim impact statements were difficult to go through, but Rosemarie felt she had to make Joan's ordeal, and their own, as real to the board as possible.

In May 1998, the court ruled that the board had set the parole bar too high. It stated that board members should not consider whether he had been rehabilitated or not, only whether there was substantial reason to believe he would commit another violent crime if released. In other words: Was he dangerous?

And that's where I came in.

5

WHAT THE PSYCH PEOPLE SAID

Over the first fifteen years of McGowan's incarceration, the case file showed, and the previously cited appeals court decision confirmed, there were at least eight psychological evaluations in addition to the initial ones conducted by Drs. Galen, Effron, and Revitch in 1974. During this fifteen-year period, McGowan appeared to be a near model prisoner, not getting into trouble and not stirring things up with other prisoners.

The first several evaluations were brief and relied mainly on self-reporting. This kind of examination of incarcerated felons is always problematic for me. When most of us see a doctor, either for a physical or mental issue, our aim is to be cured or helped, so it is definitely in our best interests to tell the truth.

This logic does not always hold up on the other side of the bars. For one thing, the felon is not seeing the psychiatrist or psychologist by choice; the visit is officially mandated. For another, as far as the felon is concerned, the encounter is not designed to help him "get better." It is to evaluate his behavior, rehabilitation, and potential dangerousness. He therefore has a vested interest not in telling the truth, but in portraying himself in the most favorable light.

On one of his court-mandated visits to a state psychiatrist following his release from Atascadero, Ed Kemper had the head of his latest victim, a fifteen-year-old girl, in the trunk of his car. During that particular interview, the psychiatrist concluded he was no longer a threat to himself or others and recommended that his juvenile record be sealed. That's why I don't trust self-reporting.

But that is what the McGowan mental health files from his incarceration amounted to. Three individual reports, generated in January 1987, October 1988, and September 1991, stated that McGowan had admitted his guilt and appeared remorseful about the crime. All three had recommended parole. On the other hand, McGowan had never reached out or attempted to express remorse to Rosemarie or anyone else in Joan's family.

On October 7, 1993, Dr. Kenneth McNiel, the principal clinical psychologist at the Adult Diagnostic and Treatment Center, met with McGowan. He was there at the request of the New Jersey state parole board. Specifically, the board wanted to assess the prisoner's "(1) likelihood of violent acting-out; (2) general personality profile; (3) presence/absence of several psychological problems; and (4) treatment program recommendations."

Dr. McNiel's findings painted a substantially different picture, not only from the three previous reports, but also from the original ones conducted by Drs. Galen, Effron, and Revitch. According to Dr. McNiel, McGowan denied "any history of sexual fantasies or behaviors toward children prior to or subsequent to his crime."

McNiel's report stated that while

> *Mr. McGowan also denied any dissociative symptoms, his discussion of the present offense was notable for brief moments in which he would blank and look away while discussing the crime, which suggested a dissociative process.*

> *[I]t was clearly difficult for him to concentrate on specific memories of his crime.*

HE CONCLUDED:

> *Mr. McGowan has made little or no progress in fully appreciating the extent of sexual deviance and violence that is apparent in his crime. Unfortunately it appears that he continues to primarily manage such negative aspects of himself through denial and repression, similar to the time of his crime.*

LIKE THE THREE PREVIOUS REPORTS, MCNIEL FOUND "NO EVIDENCE TO INDICATE Mr. McGowan is at imminent risk of violent behavior," but hedged himself by adding, "in a non-structured community setting, his ability to manage anger, rejection and feelings of sexual inadequacy remains open to question."

Taken together, these reports underscored for me the vagaries of our understanding of the human mind and motivation, or even their relationship to the physical brain. Sometimes we can look at a mental symptom and link it directly to a physical problem in the brain or nervous system, but most of the time we can't. Or, to take it one step further, sometimes we will say that a particular cruel, antisocial, or criminal action was the result of a mental or emotional disease. In other instances, we'll say that the perpetrator wasn't suffering from a mental disease per se, but had a "character disorder," and therefore is more responsible for what he did. But what is the difference between a mental disease and a character disorder? A psychiatrist reading the *DSM* can give us a definitional answer, but will it really tell us anything about the distinction?

My colleagues and I on the criminal analysis side of behavior science operate from the premise that anyone who commits a violent or predatory crime is mentally ill. This is almost ipso facto, in that "normal" people do not commit such crimes. But a mental disorder, in and of itself, does not mean the perpetrator is *insane*, which is a legal, rather than a medical, term that has to do with culpability.

There have been many attempts to define insanity over the years, but in one way or another, they all go back to the M'Naghten Rule, formulated by the British courts in the wake of one Daniel M'Naghten's attempt to assassinate British prime minister Sir Robert Peel in 1843. Shooting at point-blank range outside Peel's London house, M'Naghten instead killed the prime minister's private secretary Edward Drummond. M'Naghten, who suffered from delusions of persecution, was found not guilty by reason of insanity, and ever since, through multiple interpretations and permutations, the basic legal test of insanity in British and American courts has been whether the defendant could distinguish between right and wrong or was acting under a delusion or compulsion so strong that it negated that distinction.

Perhaps the closest we had to a genuinely insane predator was the late Richard Trenton Chase, who was convinced he needed to drink the blood of women to stay alive. When he was placed in a mental institution for the criminally insane and could no longer obtain human blood, he'd catch rabbits, bleed them, and inject their blood into his arm. When he could catch small birds, he would bite off their heads and drink blood. This was not a sadist who enjoyed inflicting pain and death on creatures smaller and weaker than him. This was an out-and-out psychotic, as opposed to a run-of-the-mill criminal sociopath. He committed suicide in his cell at age thirty by overdosing on antidepressant drugs he had saved up.

Still, there haven't been many killers like Richard Trenton Chase, and this ambiguity around insanity and mental illness highlights one of the early objectives of our project to interview killers. The conversations alone were not enough. We knew that to be truly useful, we would have to find a way to systematize our results: create distinctions that could be applied more broadly, so that there was a vocabulary that extended beyond each individual case. Back in 1980, Roy

Hazelwood, our sex crimes and interpersonal violence expert, was collaborating with me on an article about lust murder for the *FBI Law Enforcement Bulletin*. For the first time, instead of jargon borrowed from psychology, we employed a series of terms we thought would be more practical for crime investigators. We introduced concepts such as *organized, disorganized,* and *mixed* to describe behavioral presentations at crime scenes.

Roy put me in touch with Dr. Ann Burgess, with whom he had done previous research. Ann was a highly regarded author, professor of psychiatric mental health nursing at Boston College and the University of Pennsylvania School of Nursing, and associate director of nursing research for the Boston Department of Health and Hospitals. Along with Roy, she was one of the nation's leading authorities on rape and its psychological impact. Interestingly, she had recently completed a research project at Boston College involving the accuracy of predicting heart attacks in men and thought there were interesting similarities in the "reverse engineering" required for her study and what we were aiming to do.

Ann was eventually able to secure a large grant from the National Institute of Justice that allowed us to conduct a rigorous study and publish our results. Bob Ressler administered the grant and served as the NIJ liaison, and with our input, we developed a fifty-seven-page document to be filled out for each offender interview, which we called the Assessment Protocol. There were categories for modus operandi, description of the crime scene, victimology, pre- and post-offense behavior, and how they were identified and apprehended, among many other elements. Since we had already established that neither recording the interviews nor taking notes was a good idea, as soon as we finished, we would fill out the interview document, using the subject's own words, to the best of our memories.

When we finished our formalized study in 1983, we had thirty-six in-depth studies of offenders and 118 of their victims, primarily women. By this point we had enough experience and sophistication in the Behavioral Science Unit to offer profiling and case consultation on a formal basis. Bob Ressler and Roy Hazelwood continued with their teaching and research and consulted part time as their other duties allowed. I became the first full-time operational profiler and program manager of the Criminal Profiling Program, and eventually created a new unit. My first order of business was to "take the BS out of behavioral science and profiling." I renamed our group the Investigative Support Unit, or ISU. It encompassed programs in profiling, arson and bombing, the Police Executive Fellowship Program, VICAP—the national *Vi*olent *C*riminal *A*pprehension *P*rogram, which involved logging and comparing cases between jurisdictions—and coordination with other federal law enforcement agencies, including the Bureau of Alcohol, Tobacco and Firearms, and the Secret Service.

We understood and tried to make clear to potential law enforcement clients that there were certain types of crimes for which our form of criminal investigation was useful and some for which it was not. For example, a run-of-the-mill back-alley robbery or felony murder—a crime of opportunity in which quick profit was the only motive—did not lend itself to profiling or behavioral analysis. It is all too common a scenario, with a predictable profile that fits too many people to be useful. However, even in a case such as that, we might be able to suggest proactive techniques that could help flush out the offender.

On the other hand, the more psychopathology the offender demonstrates, as evidenced by the analysis of the crime, the more we can do in profiling and helping to identify the culprit. But we had to be able to undertake our analyses and consult with local investigators in a context that would use psychology but be effective in crime solving.

In 1988 Bob Ressler, Ann Burgess, and I published our findings and conclusions in book form, entitled *Sexual Homicide: Patterns and Motives*. The reception in both the academic and law enforcement communities was gratifying. But we were still working toward the goal of making our studies and research useful in a practical way to law enforcement professionals in the manner that mental health professionals use the *Diagnostic and Statistical Manual of Mental Disorders*, now in its fifth edition (*DSM-5*).

We came to realize that truly to understand an unknown subject (*UNSUB* in our parlance), you had to understand *why* and *how* he was committing a particular type of crime. And by the same token, you could classify crimes by motivation rather than simply by result or outcome. This was the challenge I tackled in the doctoral dissertation I was working on: evaluating different ways to train law enforcement officers in how to classify homicides. In other words, I was trying to present this material in such a way that it would actually help solve cases by explaining the behavioral dynamics of the crime.

The ultimate result, growing out of my dissertation research and involving some of the best minds in the FBI and law enforcement, was the *Crime Classification Manual*, published in 1992, with Ann Burgess, her husband, Allen Burgess, and Bob Ressler. By the time of the *Crime Classification Manual*'s initial publication, we already had a significant number of profiling victories under our belts, including the Atlanta Child Murders; Arthur Shawcross's murders of prostitutes in Rochester, New York; the Francine Elveson murder in New York City; the Trailside Killer in San Francisco; and the murders of Karla Brown in Illinois, Linda Dover in Georgia, Shari Faye Smith in South Carolina, and FBI employee Donna Lynn Vetter in Texas. In addition, we were also able to use profiling and behavioral science to help free wrongly convicted David Vasquez, an intellectually challenged individual who was in prison in Virginia. Though he had confessed to several murders under coercive circumstances, we were able to link the crimes to the actual killer, Timothy Spencer, who has since been tried and executed.

Looking back on it, an insanity defense like the M'Naghten Rule was one of the main reasons Ann and Allen Burgess, Bob Ressler, and I set out to create the *Crime Classification Manual*. From a criminal investigation standpoint, it didn't really matter to us whether something was a disease, a disorder, or neither. We were interested in how *behavior* indicates criminal intent and perpetration, and how that behavior correlated to the *thinking* of the perpetrator right before, during, and after the commission of the crime. Whether that behavior was disordered to the extent that it would militate against guilt (given that legally, each crime is composed of two elements—the act and the criminal intent to act) was something for the jury and judge to decide.

But these reports on McGowan's mental state made me even more uncomfortable about their role in determining his suitability for parole. If you had severe physical symptoms that definitely indicated something was very wrong and you were examined by four different doctors, each of whom came up with a different diagnosis, you would seriously question the efficacy of their diagnostic protocols. You would, unquestionably, demand a battery of tests to determine what was actually ailing you and wouldn't be satisfied until blood work and endocrine and imaging studies confirmed a specific cause for your ailment.

In most cases, though, no such tests exist for confirming the correctness of a mental diagnosis. We know the symptom—in this case, the brutal rape and murder of a seven-year-old girl—but we cannot *prove* the cause. So what concerns me the most is how accurately we can predict future *dangerousness*. It would be like the doctor saying she couldn't prove what caused a condition, but she was most interested in whether it would recur. In other words, we can only speculate, only offer an estimation or opinion. But I always start from the same premise, one that I taught throughout my years with the FBI: *Past behavior is the best predictor of future behavior.*

In 1998, five years after he first examined McGowan, and again at the request of the parole board, Dr. McNiel undertook another evaluation. Again, denying his earlier admission of rape fantasies and sexual attraction to young girls, McGowan chalked up the murder to a bad confluence of events. In McNiel's words: "The victim happened to come to his home during a moment of abject despair in which he had been actively planning to kill himself for weeks but had been unable to follow through with his suicide plans." When Joan showed up at his front door, "he became overwhelmed with unexplainable rage."

As I read over these reports in preparation for my own encounter with Joseph McGowan, one thing struck me particularly about this latest report: *a moment of abject despair in which he had been actively planning to kill himself for weeks.*

I wasn't sure whether or not he had been planning to kill himself, but from the moment I had been brought into this case and then started learning the details, my first questions had been, *Why this victim, and why then?*

Even if he was sexually drawn to little girls, and even if he was unsure of his own manhood, even if he was under the thumb of a domineering mother, what was going on in his mind at this particular time that led him to the high-risk crime of assaulting and killing a child from his own neighborhood, in his own house?

Dr. McNiel told the parole board that he considered his latest evaluation generally consistent with his earlier one, though in this later report he pointed to McGowan's "potential for dissociation at times of anger, and also the likelihood of severe sexual pathology involving pedophilia and sexual violence, which he continues to deny." He also said that McGowan had "paranoid tendencies and significant violence potential," and that, given "Mr. McGowan's continued inability to deal with the sexual aspects of his crime, it would appear that he has made very little progress in confronting the pedophilic impulses and sexual sadism that erupted in his crime. As such, he should be considered a poor risk for parole."

Okay, I said to myself. So even though Dr. McNiel considers his two reports generally consistent, and though the subject had had no serious problems in prison, while once he said he saw "no evidence to indicate Mr. McGowan is at imminent risk of violent behavior," he now sees "significant violence potential."

So what was this guy McGowan actually all about? And if I could probe deeply enough, would he show it to me?

6

RED RAGE AND WHITE RAGE

From the outside, the New Jersey State Prison at Trenton looks just what you would imagine a maximum security institution to look like: thick brownish-gray walls topped with coiled razor wire. Glassed-in guard towers stand at the corners and in the middle of the wall expanses, with the slanted tops of functional and unadorned buildings visible behind them. Even the newer part of the prison is grim-looking and fortresslike, a solid red brick structure whose narrow window slits clearly delineate the boundary between freedom and incarceration.

That morning, I had been sworn in as a deputy by a prison official and given a photo name tag that indicated I was representing the New Jersey State Parole Board. I wore my traditional dark suit to suggest my authority.

Even for someone like me, going through the outer gate of a facility like this and passing through the series of barriers that ultimately took me to the warden's office produces a sense of what Dante Alighieri must have been thinking when he posited the legend "Abandon all hope, ye who enter here," over the gate of Hell.

Before I went in to speak to McGowan, I specified several parameters that I believed, based on previous experience, would be conducive to a successful interview.

I wanted the setting to be reasonably comfortable and nonthreatening. This is no easy task in a maximum security prison, where the entire environment is intimidating and designed to be so. But within that context, I wanted somewhere that the subject would be most likely to open up. I suggested a room with no more than a desk or table and two comfortable chairs. For illumination I preferred only a table lamp—no overhead lighting. This would help make the setting subdued and relaxed.

This is very important, because in a maximum security environment, the prisoner has little freedom and I want him to feel as free in his mental association as possible—in a sense, to give him some of his power back. Then you have to keep proving yourself, not only in your knowledge of the case file and the crimes, but in your nonverbal cues. When David Berkowitz was brought into the windowless interview room in New York's Attica Correctional Facility—a room about eight by ten feet and painted a somber battleship gray—what struck me were his very blue eyes that kept darting between Bob Ressler and me as I was giving the introduction. He was trying to read our faces and gauge whether we were being sincere. I told him about the research we were conducting and that its purpose was to help law enforcement solve future cases, and possibly to help intervene with children who displayed violent tendencies.

In my research I had surmised his feelings of inadequacy. I took out a newspaper headlining his crimes and said, "David, in Wichita, Kansas, there is a killer who calls himself the BTK Strangler and he mentions you in his letters to the media and police. He wants to be powerful like you."

Berkowitz leaned back in his chair, adjusted himself into a more comfortable position, and said, "What do you want to know?"

"Everything," I said, and the interview proceeded from there.

In the prison in Trenton, I told the warden I wanted no time restrictions for the interview, nor any interruptions for food or a prison head count. We arranged ahead of time that McGowan would be fed when we concluded, even if he had missed an official mealtime.

The interview room was approximately fourteen feet square. The door was made of steel, with a twelve-by-eighteen-inch wire-reinforced window, through which the guards could check on us. The walls were cinder block, painted bluish gray. There was a small table and two comfortable chairs. The only light came from the table lamp I had requested.

McGowan had no idea ahead of time where he was being taken or why. He was brought into the room by two guards. Board chairman Andrew Consovoy, who had accompanied me to the prison, introduced me as Dr. John Douglas. He said I was there representing the parole board. I use the honorific *Dr.* only when I want to create a clinical-seeming situation. I asked the guards to remove his handcuffs, which they did before leaving the two of us alone.

McGowan and I were both in our fifties, and each about six-foot-two. I had read descriptions of him having been big but soft during his teaching days. Now his body seemed firm and muscular, after years of working out in prison. And with his gray beard, he certainly didn't look like a young high school science teacher any longer.

Everything about these interviews was orchestrated. I wanted to face the door and have him face the wall. There were two reasons for this. I didn't want him distracted, and since I didn't yet know him well, I wasn't sure how he'd react, so I wanted a clear view of the window and the guard behind it. The type of offender I interview often determines my seating decisions. When I interview assassins, for instance, I usually have to have them facing the window or door because they tend to be paranoid and will be distracted if they can't psychologically escape when stressed by the interview.

In this situation, I took a seat and positioned myself in such a way that I would be looking up at him slightly throughout the interview. I wanted to give him that one psychological edge of feeling superior to me. This was a trick I'd learned from talking to Charles Manson when Bob Ressler and I interviewed him in San Quentin. I was surprised, at five-feet-two, how short and slight he was.

As soon as Manson entered the small conference room in the main cellblock at San Quentin where Ressler and I interviewed him, he climbed up onto the back of a chair at the head of the table so he could lord over us from a superior position, just as he used to sit on top of a boulder to preach to his "family" of followers, lending him an air of natural and biblical authority. As the interview progressed, it became clear that this short, slight man who had been the illegitimate son of a sixteen-year-old prostitute, who had been partly raised by a fanatically religious aunt and a sadistic, belittling uncle who sometimes dressed him as a girl and called him a sissy, who had been in and out of group homes and reform schools and who essentially raised himself on the street before going in and out of prison for robberies, forgeries, and pimping, had developed an uncanny charisma and ability to "sell" himself to social misfits just as lost and unwanted as he was. As someone who has stared into those penetrating eyes, I can assure you that Manson's gift, his enticing aura, was real, as real as the delusional grandeur that accompanied it.

What we learned from our interview was that Manson was not a master criminal. He was a master manipulator, and he had developed that skill as a survival mechanism. He didn't fantasize about torture or murder like so many of the offenders I've confronted. He fantasized about being rich and famous as a rock star and even managed to hang out with the Beach Boys for a while.

Like other repeat criminals we had interviewed, Manson had spent a lot of his formative years in institutional settings. He told us he had been assaulted not only by other inmates, but also by counselors and guards. This had the effect of teaching him that weaker or more impressionable people were there to be used.

By the time he was released from prison in 1967, he had spent more than half of his thirty-two years in some sort of institution or custody. He made his way north to San Francisco and discovered things had changed in society. He could use his wits to participate in the culture of sex, drugs, and rock and roll, and get it all for free. His musical talent and melodic voice served him well in attracting followers. It was only a matter of time before he drifted down to the Los Angeles area and developed an "audience."

In listening to him, we realized that the horrific murders carried out in Los Angeles by his followers occurred not because Manson exerted such hypnotic control—which he did—but when he started losing control and others, particularly his lieutenant Charles "Tex" Watson, started challenging him and leading the group on adventures on his own. Manson had been predicting societal "Helter Skelter," which he picked up from the Beatles' White Album, and when he realized his acolytes had taken him seriously and murdered beautiful, nine-months-pregnant movie star Sharon Tate and four of her guests, he had to reassert himself, leading to another break-in and murder two nights later that Manson instigated but did not take part in himself.

What we learned from the Manson interview was later applied to the bureau's dealing with other cults with charismatic and manipulative leaders, such as Reverend Jim Jones's Peoples Temple in Guyana, David Koresh and the Branch Davidians in Waco, Texas, and the Freemen militia movement in Montana. The outcome is not always as we would like it, but it is important to understand the personality of those we are dealing with so we can try to predict behavior.

I had also learned from experience with assassin types like Arthur Bremer and James Earl Ray not to stare for long since it made them too uncomfortable to open up. From Bremer we learned that the target was not nearly so important as the act. Bremer had pursued President Richard Nixon before concluding that he would be too difficult to get close to and then switched his sights to Alabama governor and presidential candidate George Wallace, whom he shot and rendered paraplegic during a campaign stop at a shopping center in Laurel, Maryland, on May 15, 1972. From Ray, we got very little. He was so bound up in his paranoid fantasies that he had recanted his guilty plea in the assassination of Dr. Martin Luther King Jr., insisting that he was an unwitting dupe of a complex conspiracy to kill the civil rights icon.

Let's be honest: Any prison interview of this type begins as a mutual seduction. I am there to seduce the convict into believing that my sole purpose in being there is to help him get out. And he is there to seduce me into believing that he is worthy of getting out. It usually takes a fair amount of time to get beyond those opening positions as the pretense is gradually stripped away to reveal who we each are. My part in this case was to come across as if I was helping him think about and prepare for the big day when he would be paroled. This was not insincere on my part. I had to go into this session with a completely open mind, and with the goal to turn on the switch in his brain that would reveal his inner thoughts and fantasies.

The first two hours were devoted primarily to small talk. This amount of time is necessary to establish a natural conversational rhythm and to ensure that the subject forgets the setting and lowers his inhibitions. I told him some vague things about myself and my own background in law enforcement to build up a level of trust. I asked him about the prison environment, inquiring about what he did with his time. It was interesting to me that he pretty much kept to his own wing in the prison and seldom ventured out into the main yard, where he said he didn't feel comfortable. This was analogous to his pre-prison life, in which he felt a greater comfort level in school, where he was in control, as opposed to in the greater community, where he was socially awkward and more vulnerable.

Years later, I got to see a copy of a letter he'd written to a woman with whom he corresponded regularly. He mentioned the fact that I did not take notes, that I had the details of his file memorized, and that

I was good at putting him at ease. My purpose was to steer the conversation where I wanted it to go. He "complimented" me in his letter on listening to what he had to say, rather than merely going through a list of predetermined questions, as parole board representatives tended to do. He was right about that. I was here to learn from him by listening and getting him to reveal himself. That was my only agenda.

Gradually, I began bringing him around to the crime itself. I stayed away from sounding the least bit judgmental. It was not as if I was trying to give him the impression I was condoning his actions. I just wanted to be as factual and objective as I could, so we could recreate his thought processes at the time. He had given so many different responses to psychiatrists, psychologists, and counselors over the years that I wanted to see if I could just get him to give me the story unadorned.

I did this by creating a *This Is Your Life*–style narrative. For those too young to remember this, it was a television program during my youth in the 1950s in which host Ralph Edwards would "lure" a prominent guest into the studio with the aid of the guest's family or friends and go through his or her life for the audience, punctuated by reminiscences of people from his past. I got McGowan to tell me about himself as I guided him up to that Thursday afternoon in 1973.

I knew that he had a reputation in school for being kind of humorless and aloof, at least among the faculty. I also knew that he had been engaged to be married around this time, but that the engagement had fallen through. If the woman had rejected him, that could certainly be a precipitating stressor.

Though he said McGowan wasn't very emotionally forthcoming, Bob Carrillo had the impression that "he had a lot of emotion built up inside." He never mentioned the breakup of the engagement and Carrillo said he had never met the young lady.

"I met his girlfriend on one occasion," said Jack Meschino. "My god, she was the sweetest, most beautiful thing. She was very tiny compared to him." And perhaps a threat to his mother? Though Meschino knew at the time there had been a breakup, he never knew why, and McGowan never brought it up.

Some of his fellow teachers had planned a trip to the Caribbean over the Easter break, but McGowan had not been invited.

I asked Meschino what would have happened if Joe had asked to go along. He indicated that he probably would have been included.

Then why didn't he? Meschino said that he couldn't get himself organized and figure out how to make all of the arrangements. That was pretty consistent with a guy of his age who still lived with his mother and grandmother, and it certainly could have contributed to the ongoing frustration of the life he found himself living.

IT WAS ABOUT TWO HOURS INTO THE INTERVIEW WHEN I SAID, "I WANT TO KNOW IN your own words what it was like twenty-five years ago. How did this all happen to get you here?" I purposely avoid loaded or descriptive words like *kill*, *assault*, or *murder*, nor did I refer to Joan as his "victim" or a child. "That girl—Joan—did you know her?"

"Well, I'd seen her in the neighborhood," McGowan replied. His affect was flat, his tone matter-of-fact.

"Had she come to the door to sell you the Girl Scout cookies?"

He said he thought his mother had ordered them from her. A newspaper article quoting a former FBI agent said they found more than a hundred empty boxes of Girl Scout cookies in the house.

"Let's go back to the moment she came to the door," I said. "Tell me what happened, step by step, from that point on."

It was almost like a metamorphosis. McGowan's whole demeanor transformed. Even his physical appearance seemed to change before my eyes. His eyes were unfocused as he looked beyond me and stared toward the vacant cinder-block wall. I could tell he was looking completely inward—back a quarter of a century. I could sense he was clicking back to the one story that had never left his mind.

The front door was open to the mild spring weather, and from the lower level of the bi-level home, McGowan said he saw her through the screen door standing on the landing. She said she was there to deliver two boxes of Girl Scout cookies and to collect two dollars. He wanted to get her downstairs to the level where he lived, away from his grandmother, who was either sleeping or watching television upstairs.

That was why he said he had only a twenty-dollar bill and a one-dollar bill and would have to go get the proper change—to get her downstairs with him. The story about being embarrassed over not having the exact amount was all crap to give something innocuous to the psychiatrists.

This is what I was hoping would happen during the interview. I've had similar experiences with other sex offenders: Once you're able to flip their on switch, they don't shut up. When Ressler and I interviewed Monte Rissell, he recounted driving back to the parking lot of his apartment complex in Alexandria, Virginia, seeing a woman about to get out of her car, and forcing her at gunpoint into a secluded area. After that, she tried to run away. He pursued her into a ravine, grabbed her, and vividly described banging her head against the side of a rock and holding her head under the flowing water of a stream, as if he were watching a movie.

My goal was to turn on the "DVD" in McGowan's brain that had recorded the homicide. After a quarter of a century behind bars, McGowan recalled every specific detail of that Thursday afternoon. It was like having a friend talk about a terrific movie he's seen. But in this case, McGowan was the scriptwriter, the producer, the director, and the lead actor. He got to exercise the three aspirations of nearly every predator: manipulation, domination, and control.

Without looking directly at me, he described luring Joan into the lower level and his bedroom, ordering her to remove her clothes, and sexually assaulting her.

I asked him if he had penetrated her. Just with his finger, he insisted.

Then how did semen get into her vagina? It was on his fingers after he ejaculated, he said.

In a focused and excited state with no indication of any remorse, he described how he picked her up by her ankles, swung her around, and slammed her head to the floor, fracturing her skull. The account he gave was not much different from what he had told detectives and psychiatrists previously. He never even attempted to fake empathy, as other offenders had when I'd interviewed them. What struck me was not the facts, but the obvious *intentionality*.

It was hot as hell outside, but the interview room was very cold. In fact, I was trying to keep myself from shivering as I sat there, even though I was wearing a wool suit. McGowan, on the other hand, had begun perspiring profusely, just as he had described feeling after the attack. He was looking away from me in a trancelike state, breathing heavily, and his prison shirt was drenched with sweat. I could see his chest muscles trembling.

I thought immediately of the quotation I had just read from Dr. Galen's initial report, after McGowan had beaten and strangled Joan: "she stopped struggling . . . just sort of lay there. I got dressed. I had been sweating so violently." In his head and in his physical being, he was right back in the act.

"It's pretty hard to strangle someone, isn't it?" I asked. "Even a very young person."

"Yeah," he readily agreed. "I didn't realize it would take so much strength."

"So what did you do?"

"Well, I turned around and positioned myself behind her." I took that to mean he came around to where her head lay on the floor.

"And how long did you choke her?"

"Until I thought she was dead."

"Then what?"

"I went out to get some bags to stuff her in and put her clothes in, and when I got back, she was twitching."

Now, this was not a blitz-style attack resulting from a momentary surge of uncontrollable rage. He did not suddenly come to his senses and say to himself, *Oh, my god! What have I done?* When he saw that she was still twitching, his only thought was to choke her again to make sure she was finally dead. It was almost as if he had murdered her again.

The one time I remember McGowan looking right at me during this entire description was when he said, "John, when I heard the knock and looked up through the screen door and saw who was there, I knew I was going to kill her.

"I can feel two different kinds of rage," he went on. "Red rage bothers me, but I can turn my head, focus, and control it, like when someone cuts me off in traffic or I have a conflict at school. But white rage I can't control."

"And is that what you were feeling when Joan came to your house?"

"Yeah," he said. "That's right." Our eyes were still locked.

So he killed Joan while seized with white rage. The fact that he even had a name for it told me he had experienced it before and most likely would again. But he didn't kill her right there at the front door no matter what was raging within him. At that moment, a methodical plan had formed instantly in his mind as to how he was going to get her where he wanted in order to be able to do what he wanted to do.

I pointed this out and said to him, "You're not a psychotic, even though you tried to show dissociation from the crime. What I see is pretty logical, rational behavior." He didn't argue with me.

As for the psychiatric assertion that "The killing was the consequence of an additional upset and failure due to premature ejaculation": *wrong again.* The killing was the result of a combination of displaced rage, sexual excitement at his momentary power over another human being, and the very practical consideration of leaving no witness to what already was an unspeakable crime.

He had admitted as much in his interview with Dr. Galen when he said, "If I let her go, my whole life was gone. All I could think of was just to get rid of her." I wondered if anyone before me had ever taken the time to correlate all of McGowan's statements together.

But I didn't minimize the killing as merely a practical consideration to avoid being identified and caught. This was not a question of simply going too far. The way McGowan described it, I could tell that the gratification and emotional fulfillment of the act went through to the brutal murder itself and his ability to destroy something or someone.

At one point I asked him, "If I could have captured one 'video image' of your face while you were doing this, what would it show me?"

Obligingly, his face contorted into what I would characterize as an intense, malevolently satisfied grimace.

Even the way he described the body disposal was logical and methodical, as opposed to panicked or hurried. He got some towels and cleaning materials to wipe up the blood, trying to get this potential forensic evidence out of there so that his mother wouldn't see it and in case the police searched the premises. He wrapped the body in a carpet and drove some distance to a setting he was familiar with to dump it. Then he went back home and acted as if nothing had happened. Joining the neighborhood search for Joan was a conscious means of covering his tracks.

Most cons, knowing they potentially have something to gain from the interviewer's good report, will try to BS to one extent or another. There was none of that with McGowan. I think that was because I was well prepared, and he was intelligent enough to understand that it wouldn't work. His goal in agreeing to talk to me was to improve his chances of parole, and he knew that my catching him up in a lie was not going to cut it. And the whole point of my approach was to make the subject "comfortable" enough to let me know what he was actually thinking and feeling.

What did surprise me was that he didn't even try to express any sorrow for what he had done or feeling for Joan's family. He was certainly sorry that the whole thing had happened—I was not the first one to hear that sentiment—but there was no sense of emotional comprehension of what he had taken away from this little girl, all those who loved her and all those who had come into her brief life. The impression I received was that he was telling me that since he couldn't bring this young girl back to life, he just had to move on, and everyone should understand that. The murder was simply a fact of life to him, as if he had had cancer or a heart attack, and now the doctors were determining if he was well enough to leave the hospital and resume normal life.

Every violent crime is a scene acted out between two or more participants. And when the offender and the victim are up close and personal—as opposed to, say, a bombing, a poisoning, arson, or a sniper attack—a trained crime analyst can gain a tremendous amount of information on what is actually going on in the mind of the criminal by observing his behavior, even if he says little or nothing. As I listened to each detail of the deadly encounter between Joseph McGowan and Joan D'Alessandro, I focused not only on the fact of the violence and sexual assault but on the *way it was done.*

The fixation was with the act itself, not the personality or specificity of the victim. I didn't doubt McGowan had some pedophilic tendencies, as he'd suggested in some of his previous interviews. That would certainly go along with his lack of social sophistication and his deep-seated feelings of inadequacy. But he had no prior sexual offenses on his record, even minor ones, and there was nothing in the file about search warrants turning up any child pornography or fantasy writing about children. If he fantasized, I thought it would be about an adult woman. And the only real fantasy here was a fantasy of power.

What became clear to me as he talked was that this crime, primarily an act of rage that escalated quickly from "red" to "white," was triggered by *something*, some precipitating event, though I wasn't sure at this point what it was. My guess—based on his living situation, his inability to see his engagement through to marriage, and my extensive knowledge of other sexual predators—was that it had something to do with his mother. Still, it gnawed at me that I couldn't pinpoint exactly what had set him off.

As I listened to McGowan, I realized that what had gotten him off was the *act* rather than the *victim*. All the people who had interviewed him before were barking up the wrong tree if they were trying to get him to talk about his pedophilia. There was nothing either obsessive or particularly pleasurable in his description of the rape of a seven-year-old girl. Not only that, this was a child he had seen frequently in the neighborhood. He had never previously attempted to single her out for special attention, to befriend her, seduce her, or groom her. The violence, the sexual degradation, the murder—these were all manifestations of raw rage. No fantasy scenario was being acted out, not even a sexually sadistic one. What was significant about this particular victim of opportunity was that she was small and vulnerable. If it had been someone who McGowan saw could put up a good fight, no crime would take place.

McGowan slowly returned from his reverie excursion into the past. While describing the particulars of the crime, he had been focused and trembling. Now he was calm, no longer sweating. He had relived a battle he had fought and won, unlike so many others in his life.

We talked about his fondness for guns, another obvious psychological compensation. I asked, "If you were angry and you went out with an AK-47 to a shopping mall, who would you kill?" I was curious not only about his response, but also to see if he would even accept the premise of the question. "Whom would you be gunning for? Schoolchildren, teachers, police officers?"

"Anyone," he replied.

This was significant. Not only did he not deny the possibility that something like this could happen, but he essentially told me that his rage was generalized and indiscriminate.

We began to talk about his possible release from prison. At one point I asked, "Joe, where are you planning to go when you get out?"

I was careful to say *when*, not *if*. I wanted to keep the conversation as positive as I could so he would be candid with me.

He told me he was going to New York to meet up with another ex-con who was an electrician. He had promised McGowan a job as his assistant. I told him that I was raised in New York and returned frequently, and that he would be shocked to see how expensive it had become to live there.

He glanced furtively over his shoulder at the door to make sure the guards couldn't hear our conversation. "John," he said in a conspiratorial whisper, "I have money."

"What kind of money could you have, being in a place like this for the past twenty-five years?" I responded. "It can't be from making license plates."

In a low voice he said that when his grandmother and mother passed away, he received a fairly substantial chunk from their life insurance policies and the sale of the house. He said it was in a bank out of state. Hundreds of thousands of dollars.

"Why's that?" I asked.

He whispered, "I don't want the victim's family to be able to get at the money."

What I was thinking to myself was: *This guy just won't let up. He has absolutely no regard for the people he's hurt and whose lives he has changed forever.*

What I said was "You know, you're a pretty smart guy, Joe, the way you've figured it all out. I think you'd do really well in New York!" This was both necessary to maintain my rapport with the subject and not a lie. I did think this was pretty smart and resourceful and that he would be able to figure out his way through a place as intense as New York. I just didn't say how appalled I was at his scheme. Just as in negotiating a high-level business deal, you have to know when to speak up and when to shut up, as difficult as it may be.

We later found out that an investigator had looked into Genevieve

McGowan's will and the money trail on Rosemarie's behalf. She had sold her Hillsdale home shortly after the murder and moved to Villas, New Jersey. When she sold that house, she moved in with her niece in Wisconsin for a while, and then went to live at a Franciscan care center. She died in April 1992. Her will set up various trusts, one of them benefiting a niece.

The will was probated in Wisconsin since Genevieve was living there. Jim discovered that all funds had been disbursed and there was nothing left for the D'Alessandros to touch. Apparently, Genevieve had anticipated claims on anything coming directly to Joe, so she had protected her assets from any claims arising out of a wrongful death suit by making disbursements throughout the years. She had instructed the niece to take care of Joe and give him whatever he needed, without actually having him legally control the money. This is what he must have meant when he told me the money was held and protected out of state.

The sentiments behind these arrangements reminded me of Genevieve's comments to her church acquaintance that she hated Rosemarie and blamed her for all of her and Joe's troubles. One of the hallmarks of narcissistic, borderline, and sociopathic personalities is the unwillingness to assume personal responsibility for anything. It is always someone else's fault.

By the time the conversation wound down, five or six hours had passed. Neither one of us had eaten or left to go to the bathroom. But I did have a pretty good idea now of what made Joseph McGowan tick. He sensed this as well, although his perspective turned out to be somewhat different from mine. In the letter to his female pen pal, McGowan expressed optimism about parole because he thought that he had had a good interview with me and that I had understood him. I did understand him, and I went in with an open mind, not about what he had done—that was beyond debate—but whether he still might be dangerous. Where he was in error was interpreting my nonjudgmental attitude and demeanor as empathy or acceptance.

That is where almost all of these guys go wrong—they can process other people only through their own self-centered emotional filters. It's always all about them, and they cannot comprehend that my true empathy for them is about equal to what they showed for their victims.

As we concluded the interview, I shook McGowan's hand and thanked him for talking with me. I wished him good luck and tried to give no indication of my personal feelings about him or my recommendations to the parole board.

7

THE BOTTOM LINE

The next morning I went before the full New Jersey Parole Board. Most parole decisions were made by only two or three members, but since this was a high-profile case that would be controversial whichever way the decision went, Andrew Consovoy wanted the entire panel on the record.

We met in a conference room at the prison. I think there were about ten or twelve people in the room; the collective experience included legal, psychological, and police work. Consovoy introduced me and asked me to give a brief overview of my background in profiling and investigation. I recounted the origins of the FBI's behavioral science and profiling programs and explained that my doctoral project concerned teaching police officers and detectives how to classify crimes of violence.

I told them that I try to remain objective going into each case and did not read all of the reports until the day before the interview.

"The basic thesis, the basic premise of my approach," I stated, "is that to understand the artist, you must look at the artwork." Likewise, I clarified, to understand the violent criminal, you must look at the crime.

No sense mincing words at the beginning, I decided. They might as well know where I'm coming from. "And I never could understand," I went on, "people in a position of making decisions relative to probation, parole, sentencing, and treatment—if you didn't have that information, if you don't understand what it's telling you and you don't understand the person sitting across the table from you, if you believe that person is telling the truth but you're relying only on self-reporting, you're going to have the wool pulled over your eyes."

For example, if you're trying to evaluate a convicted rapist, you need to review the police interview of the victim about what he did and said during the assault to understand which of the five distinct rapist typologies he fits. If he has murdered the victim, that obviously tells you much of what you need to know right there.

Before the meeting began, Consovoy had told me that a number of members of the board were interested in the pedophile angle. If McGowan was paroled, should he be classified as a sex offender?

I said to the board, "I was curious to see whether we were looking at a traditional type of child molester, a pedophile—looking for whether or not this is a 'preferential' type of offender, looking for a specific type of victim, or is it a 'situational' offender? What I mean by that term is, whoever crosses the path of this type of offender could potentially be a victim. So what we have to do is evaluate the risk level for the offender and the risk level for the victim.

"As far as victimology, the child goes from a low risk inside her home and yard, to a moderate risk in the neighborhood, to a high risk once she steps inside a house she's never been in before."

For the offender, the risk for the act itself is low. There was little doubt he could do whatever he wanted to a seven-year-old. But the risk of identification was high. The crime was committed in the offender's and victim's neighborhood, the victim could identify him if left alive, and there was a reasonable certainty that one of the victim's parents or someone else would know where she had gone. Therefore, it should have gone through the offender's mind that it would be just a matter of time before the investigation would be directed at him.

When we did the research behind *Sexual Homicide: Patterns and Motives* and the *Crime Classification Manual*, we began dividing predators by organized, disorganized, and mixed presentations. I explained that there would be several possible reasons for a disorganized offender to undertake such a high-risk crime. These would include youthfulness and inexperience, judgment or impulse control mediated by drugs or alcohol, loss of control of the situation, or mental defect. McGowan was none of those.

It wasn't as if he woke up that morning and said to himself, *I'm going to wait for someone to knock on my door and then I'm going to kill him or her.* But this crime, while clearly opportunistic, was organized. It showed a logical thinking process. This is something many people, even law enforcement personnel, have a difficult time understanding: If the crime itself is so illogical, how can the process of carrying it out be organized and methodical? In other words, how does someone like Joseph McGowan—who is intelligent, educated, and respectable, heavily invested in society's regard through his status as a public school teacher—take such an action that jeopardizes everything he has worked for and considers important? How can this happen?

The answer is that it does happen, and usually because the impulse for the act is triggered by something more powerful than the rational thinking process. In this case, that something appeared to be this ongoing and overriding feeling of inadequacy and low self-esteem, coupled with the specific cause of overwhelming anger, looking to vent itself in explosive rage.

I explained that between the ages of approximately twenty-five and thirty-five, certain individuals—and this applies to men overwhelmingly more than to women—realize that they're not going to amount to what they think they should. Even though Joseph McGowan had a good job, he would be forced to face the fact that he was still living with his mother and had not become the man he wanted to be. The anger begins to build, and these types of feelings don't just burn out; quite the contrary, they get worse as these individuals accept the fact that they are not going to meet their own goals and expectations.

This was a crime of anger. Sexual assault was merely one of the weapons. It was justified in his own mind, at least in the moment, by what he perceived had been done to him by others. And though McGowan would not have thought of it in this analytical fashion, Joan became the representative and surrogate for all those others. For a person with criminal tendencies who perceives he has little power or control over his own life, murder embodies the ultimate power. For that one brief moment or however long he can extend the experience, he has ultimate control of the immediate world around him. My sense of it, I reported to the parole board, was that McGowan had never experienced this kind of feeling before, and when it occurred, it was all-encompassing, hypnotic, transcendent.

"He gets her down to the basement bedroom. He makes her strip down. He prematurely ejaculates. He's excited, but not because he's going to fulfill his fantasy and have sex with a young girl. He's excited because of his power, because he wants to kill and he's going to—that's what causes the erection. And then his anger is intensified even more when he loses control—not of his victim, but of his own erectile function."

Another concept difficult for many people to understand is that an offender can be sexually aroused by something that does not appear to have any direct connection with sex. When we interviewed Son of Sam killer David Berkowitz, he told us that when he used to set fires and then watch the fire department arrive on the scene, he would masturbate. For this nobody, the power to control great forces—fire itself and the human force of the fire department, plus all of the curious spectators—was a sexual act. Likewise, he told us, during his killing spree, he would return to the locations where he had shot young couples, absorb the atmosphere, then go back home and masturbate, reliving the fantasy of the power of his kills.

Dennis Rader, the BTK Strangler, admitted to me that he loved to drive around past the houses of his victims. He considered those houses trophies as well and gloried in the fact that no one knew his secrets. He said he stayed away from his victims' memorial services and graves, much as he might like to be there, for fear of surveillance. Instead, he cut their obituaries out of the newspaper and read them over and over again. Indeed, this sense of power that murder creates is incredibly seductive to these killers.

One of the board members noted that McGowan had said in previous statements that Joan had followed his order to remove her own clothing and that she was not crying or protesting in any way during that time.

"I find that hard to believe," I said. "I think she actually did not obey every command, which meant he lost control." It was inconceivable to me that this young girl would not be terrified and crying. And my reading of the medical examiner's report certainly indicated some sort of struggle. Rosemarie had said on several occasions that her daughter would not have taken any assault passively. It was my strong opinion that in his recounting, McGowan was simply trying to minimize the cruelty of his crime.

While the medical examiner's report offered the opinion that Joan had suffered penile rape, I believed from what I gathered in the interview that McGowan had only digitally penetrated her, and that accounted for the ruptured hymen.

I brought up his previously proffered excuse of not having the right change when Joan came to the door. "He didn't have the right change? As motivation for the crime, that's a crock. He was just looking for some justification. He had intent to kill from the beginning, whether he had the right change or not. As for any doctor saying that this never would have happened if he'd had the right change, that's ludicrous!" I told the board members about "red rage" and "white rage."

The board members were further surprised when I told them about the money he'd stashed out of state and his plans for what to do if granted his freedom. I had asked McGowan how he thought he'd do in the outside world after twenty-five years in prison, many of them in maximum security.

"If I can survive here, I can survive anywhere," he'd replied. But while no one doubts the harshness of prison or what it takes to get by in the Big House, I noted that life in prison is quite different from life on the outside. For all its possible horrors, prison is a very controlled, highly structured environment. McGowan was getting three meals a day, psychiatric medication, and constant supervision. Violent offenders who can't function properly in the outside world often do well under these conditions. I told the board that while you certainly would not consider paroling a troublesome prisoner, from my experience, the fact that one is a cooperative or model prisoner has little predictive value in determining how dangerous one is beyond the prison walls.

We talked about the nun from Connecticut with whom McGowan had been communicating and her offer to get him placed in a halfway house. I pointed out that he had acknowledged he would want people watching him the way they had been in prison and didn't know what would happen if he didn't have adequate supervision. He would never be allowed back into teaching and would have trouble interacting with those he considered intellectually inferior, so he would need some kind of solitary job. This would be a lot of needles to thread.

McGowan had told previous mental health practitioners that he remembered having been aroused by seeing his twelve-year-old cousin in a short nightie and noticing her delicate pubic hair peeking out from her panties. But I said he also would have been aroused by a movie star like Raquel Welch or any other adult attractive woman in the same attire. The difference is that he wouldn't have the courage to approach her as he would a twelve-year-old girl. This says more about his level of social sophistication than it does any pronounced pedophilia.

"If this had been a premeditated sexual crime," I stated, "he would have driven around the neighborhood or adjoining neighborhoods looking for someone unknown to him, someone harder to trace. Instead, he attacks the first vulnerable person who comes to his door." I said I wouldn't even classify him as a sexual predator. He didn't get off on little girls so much as he got off on domination and control.

"I don't look at him as a classic child molester," I said, "because if he gets out, if there are any roadblocks, any stumbling blocks, any frustration along the way, don't expect necessarily that he is going to go out and molest another seven-year-old. The victim may change, but the anger is still there."

I was reminded of the case of Jack Henry Abbott, a killer and repeat felon who had spent much of his adult life behind bars. When Abbott heard that author Norman Mailer was writing a book about Gary Gilmore of Utah, the first person executed in the United States after the Supreme Court reinstated the death penalty in 1976, he offered to provide Mailer with realistic descriptions of prison life. Based on the insight and raw literary talent evidenced by the letters, Mailer helped the convict publish a memoir entitled *In the Belly of the Beast*. The book received positive reviews and a lot of attention and was used by Mailer and other prominent people to support a move for Abbott's parole, concluding that someone who showed that degree of insight and sensitivity in his writing was rehabilitated.

Despite the misgivings of prison officials, Abbott was granted parole in 1981 and came to live in New York, where Mailer and his family endeavored to find him work and reorient him to life on the outside.

Six weeks after his release, Abbott was having dinner with two women at a café in Greenwich Village. When Abbott got up to go to the bathroom, he got into an argument with a waiter named Richard Adnan, an aspiring actor and playwright whose father-in-law owned the café. They ended up outside, where Abbott stabbed Adnan to death.

Mark became friends with Mailer in the great writer's later years, and he told Mark that the entire Abbott episode was one of the greatest regrets of his life. I made clear my concerns of a similar scenario in McGowan's case if something happened that tripped his white rage.

Hearing this, Consovoy said, "Our responsibility is to determine his level of threat." He paused, looked at me, and asked, "If you were sitting on the parole board, would you release him?"

"No," I replied. "I don't know when he's going to commit his crime. I don't know whether it's going to be a year, five years, or ten years. But when the situation presents itself, when life presents him with any stressor—loss of a job, rejection by a woman, rejection by a community that doesn't want him living among them—he can lash out again. I look at his personality as a ticking time bomb ready to go off if things don't go his way."

I brought up his response to my hypothetical scenario of going into a shopping mall with an AK-47.

"He simply can't cope with stress. That's why he broke down on interrogation."

And it wouldn't have to be an earthshaking incitement for him to turn again to violence, I pointed out. "For instance, someone cuts in front of him in line at the supermarket?" one of the board members offered.

"He might go out to the parking lot and wait in his car," I suggested. "He has this anxiety, panic attack reaction. So to overcome this, what he does is psych himself up, get angry as hell. He goes back into the store and he confronts this person. And if she isn't nice, he explodes."

I noted that now that he is older, his M.O. could change and he might switch to other types of victims. I had seen this often. If he decides to target prostitutes, his social inability would cease to be much of a problem. All he would need is a vehicle. It would be up to the prostitute to approach him and begin the conversation, rather than the other way around. All he would have to do is tell her to get in the car.

I cited the case of Arthur Shawcross, known as the Genesee River Killer, in Rochester, New York. He killed two children—a boy and a girl. He was sentenced to twenty-five years to life and was paroled on good behavior after fifteen years. Then he started targeting prostitutes and murdered twelve women before he was caught. The details changed and so did the victims, but his prey still consisted of vulnerable, easy-to-approach individuals. I didn't want to see a repeat of Shawcross here.

"When there is pressure, when there is a crisis, you'd have to be watching him twenty-four hours a day."

Ultimately, we moved on to a subject that naturally comes up in these situations. "I understand you're not supposed to address rehabilitation in this case," I said.

"You can, we can't," Consovoy clarified.

Then I would. "When you're dealing with offenders like this, the word *rehabilitation* should never be used. Because he's never been *habilitated*. Get him back? Get him back to what?"

"So, John, do you see any appreciable difference between him now and when he first went to prison?" Consovoy asked.

"I don't see it," I said. This was a crime of anger, a crime of power. It was not about sex for its own sake. As I've noted about so many predators, this was about manipulation, domination, and control.

"He'll put on a good front for you, but he's like an iceberg—you're only seeing the ten percent above the surface. And you may see changes in the way he responds, because he's intelligent and the whole parole process has been an education to him. He knows all your tests. He knows what you're looking for.

"But all you've done is put his physical body on ice for the past twenty-five years. You have not changed what's in his head—the sexual response to the power of violence."

This was also the case with Shawcross. When my unit was called in to assist on the hunt for the Genesee River Killer in the late 1980s, profiler Gregg McCrary created what turned out to be a highly accurate profile and strategy that helped lead to the apprehension of Shawcross. The one element Gregg got wrong was the suspect's age. He underestimated it by about fifteen years. Those fifteen years in prison had merely been a holding pattern for him; he resumed his previous life and attitude as soon as he was released.

"If [McGowan] re-offends, it's likely to be against a person or persons near at hand," I told the board. "Bottom line: I would not want this person living on my block or in my community."

“SUBSTANTIAL LIKELIHOOD”

I had learned a lot about, and from, Joseph McGowan. But I remained dissatisfied with one element of my analysis. One piece of the puzzle remained missing. I felt sure there had to have been one precipitating stressor or inciting incident that had set him off to kill an innocent child. It didn’t mean that something had set him off and made him “decide” to rape and kill a little girl, or that the crime wasn’t a spontaneous decision in which motive, means, and opportunity suddenly converge to make it possible to fulfill a dark, already existing desire. But I felt sure that something had “conditioned” him to act around the time he did.

Was it something on the job—a fight with a colleague, an affair or rejection from a student that he had been unwilling to mention to me? Did it have something to do with being dumped by his fiancée? That certainly could have triggered rage. I felt reasonably sure his not being included in the Easter-break trip with his colleagues had something to do with it. But after studying the case file and talking to him for an extended period, I thought that there had to be something more significant going on.

We always wanted our prison interviews to be open-ended and far-ranging, because you never know which element or line of questioning is going to lead to something valuable. But in certain murder cases and other violent crimes, there is one key question that mystifies investigators, the solution of which would provide a major clue.

The term *serial killer* designates a predator who murders repeatedly and with a certain general periodicity. And after each crime, there is a "cooling-off" period. If the killer stops without being caught, it is almost always for one of three reasons: He has died; he has been arrested for an unrelated crime and is in prison; or he hasn't actually stopped, but has merely moved to another area and law enforcement has not connected his new crimes with the older ones. But with the BTK Strangler in Wichita, there were long periods between crimes and then we would hear from him again, either with a new murder or a written communication to the media or police bragging about a previous one and providing evidence he had done it. The "credit" for these horrific crimes was clearly so critical to his ego that we couldn't figure out why he lay dormant for so long.

The case of the BTK Strangler of Wichita, Kansas, began early in my FBI career, when on Tuesday, January 15, 1974, Charlie Otero, two weeks short of his sixteenth birthday, walked home from school with his fourteen-year-old brother, Danny, and his thirteen-year-old sister, Carmen, to find that their mother and father, thirty-four-year-old Julie and thirty-eight-year-old Joseph, had been bound, gagged, and brutally strangled and stabbed. When police arrived at the scene, they discovered the body of nine-year-old Joey, bound hands and feet, lying on his side, strangled, in the bedroom he shared with Danny. In the basement they found eleven-year-old Josie's body, hanging by a noose from an overhead pipe, her hands tied tightly behind her back. Like the others, she was completely bound with cord. Her mouth was gagged with a towel, and like her father, her swollen tongue protruded from her mouth above the gag. She wore a pale blue T-shirt and her panties dangled around her ankles. On one leg was a sticky substance that appeared to be semen. The killer had masturbated over her as he watched her die, or afterward.

This was the beginning of a brutal and sadistic series of killings in the Wichita area that went on for seventeen years, terrorized the community for more than thirty, and claimed the lives of at least ten victims.

Ten months after the Otero killings, a local newspaper received an anonymous call directing authorities to a mocking letter in a library book claiming credit for the murders, promising more, and concluding: "The code words for me will be . . . Bind them, torture them, kill them, B.T.K., you see he at it again. They will be on the next victim."

The boastful communications with the police and media would continue, as if this hunger for attention and credit was just as important as the torture-murders themselves.

I was asked by the Wichita police department to do an investigative analysis just around the time the operational profiling program at Quantico was getting up to speed. We took part in a major case consultation with members of the BTK task force ten years after the UNSUB had committed his first murder. And by the time the case finally reached resolution, I had already been retired from the bureau for ten years. The viciousness of the crimes, both physically and psychologically, continues to haunt just about everyone who became involved in the hunt.

He chose his targets both through opportunity and planning, sometimes following people he had seen in his endless driving around town or encountered in his job as a municipal enforcement officer, attending to such weighty matters as overgrown lawns and stray dogs. After the Otero family murders, his victims ranged in age from twenty-one-year-old Kathryn Bright, up to sixty-two-year-old Dolores Davis. And he was absolutely heartless. He strangled twenty-four-year-old Shirley Vian to death in her bedroom within the hearing of her young children, whom he had locked in the bathroom.

But from an investigative standpoint, the strangest and most perplexing aspect of the case was the temporal irregularity of the crimes.

There were the five murders in 1974, two more in 1977, one each in 1985 and 1986, then one more in 1991. Even if BTK had died or been put away for an unrelated crime after that, it didn't explain the years-long gaps before he resumed his hideous activities. This would hold especially true for a serial killer so interested in bragging about his work and insisting on his own place in the firmament of media infamy.

BTK's downfall did not come until 2005, fourteen years after his last known kill. Not content to rest on his murderous laurels, he pushed one step too far by sending a computer disk to a local television station that police technicians were able to trace through its metadata to a local church to its last user, "Dennis." An Internet search listed a Dennis Rader as president of the church council, and his black Jeep Cherokee matched the description of a vehicle leaving the scene where one of BTK's communications had been left.

He turned out to be a family man with a son and a daughter. The cowardly Rader agreed to a plea deal without the possibility of parole to avoid the death sentence he had—much more sadistically—imposed on his innocent victims. It was in the maximum security El Dorado Correctional Facility in Kansas that I had the opportunity to confront BTK directly.

After talking to Rader and studying his case, I had an idea of what caused the killing gap between Nancy Fox in 1977 and Marine Hedge in 1985. It had to be his wife, Paula, I speculated. She must have found something or come in on him doing something.

Rader confirmed that sometime in fall 1978, Paula walked into their bedroom and found Dennis wearing a dress, with a rope around his neck, hanging himself from the bathroom door. Cross-dressing and autoerotic asphyxiation were two of his favorite masturbatory activities. The dress was not hers, so it presumably came from one of the many homes he had broken into over the years. The mild, sheltered Paula could not believe what she was seeing. She'd never even heard of this kind of thing before.

She told him he needed help, but she had no idea where he should go. The whole thing was just too embarrassing to talk about. Instead, after a few days of brooding, she called the VA hospital where she'd once worked as a bookkeeper and asked to speak anonymously to a therapist. She said "a friend" had a husband whom she'd caught wearing women's clothing and trying to hang himself. The therapist recommended several self-help books, all of which Paula bought and gave to Dennis.

For his part, Rader pled that this was a psychological problem he'd been fighting for years and promised never to do it again. Terrified that any action by Paula might lead investigators to him—she had once offhandedly commented that his handwriting looked like that of BTK in pages the police had published—he decided he'd better lie low and try to stay away from the autoeroticism, at least in the house.

Apparently, this worked for about two years. Then, in 1980, Paula once again came into the bedroom to find Dennis with a rope around his neck. This time she wasn't concerned for his health so much as "mad as a goddamned hornet." He'd never seen his wife so stirred up and angry, and it scared him. She said if he ever did this again, she would leave him. If she went public with what she'd seen, and there was another BTK murder, how hard would it be for authorities to put two and two together?

This was a new insight for me into why a serial predator might cease his crimes on his own, and it gave me a new understanding of the types of people I'd spent my professional life hunting. Ironically, despite Paula's fears for her husband the first time she caught him acting out his sexual obsession, this layoff proved to me Rader's sanity and rationality, rather than the opposite. While I do believe he and all vicious killers and rapists have varying degrees of mental illness, the fact that he could choose to stop, even temporarily, for his own survival shows a high level of prospective thinking and executive function. Of course, with his drawings, S&M pornography, self-bondage, transvestism, autoeroticism, and crime scene souvenirs, not to mention his vivid imagination, few serial killers are as well equipped as

Rader to substitute for the thrill of the actual act. I have little doubt that he will continue to fantasize about tying up, taunting, and watching women and girls die at his hand as long as he lives.

A different kind of mystery surrounded the capture of master con man, bank robber, jewel thief, and aircraft hijacker Garrett Brock Trapnell. His most spectacular crime took place on January 28, 1972, when he hijacked TWA Flight 2, a Boeing 707 jet, from Los Angeles to New York, over Chicago, using a .45-caliber handgun he had smuggled onto the plane in a fake plaster cast on his arm. He demanded more than $300,000 in cash, an official pardon from President Richard Nixon, and the release of professor and activist Dr. Angela Davis, who had been sentenced to prison on conspiracy charges for supplying weapons to a defendant who gained control of a Marin County, California, courtroom and took hostages. The judge and three other men were killed. Many considered Davis's conviction politically motivated, and her conviction was later overturned.

With the plane on the tarmac in New York for refueling and a change of crew, two FBI agents disguised as crew members boarded, shot Trapnell in the left shoulder and arm, and captured him. After one five-week trial ended in a hung jury, he was convicted of air piracy and sentenced to two life terms plus eleven years in prison.

Even after his conviction, Trapnell wasn't through with spectacular stunts. He somehow convinced his friend Barbara Ann Oswald, whom he had met while she was doing a graduate study program on prisoners, to hijack a charter helicopter in St. Louis on May 24, 1978, and force the pilot to land in the prison yard at Marion to rescue him. During the landing, the helicopter pilot was able to wrestle away Oswald's gun and kill her.

On December 21 of the same year, Oswald's seventeen-year-old daughter Robin attempted to hijack TWA Flight 541 from Los Angeles to New York, demanding that Trapnell be freed or she would detonate a dynamite pack strapped to her torso. FBI negotiators were able to talk her down without any injuries, and the dynamite pack turned out to be railroad flares wired to a doorbell. Mother *and* daughter? I'd seldom seen an offender who had this amazing power over certain types of impressionable people. Charles Manson was the only other one who came readily to mind.

But what intrigued me most about Trapnell was his demand to free Angela Davis during the airplane hijacking. Political airplane hijackings at that time were far from unknown, and the planes often ended up in either Cuba or Algeria before being returned. But from everything I'd been able to learn about him, Trapnell had no strong political commitments. His only strong commitment was to himself and his own enrichment. So why would he make such a strong point of this demand, even as he was being led away by federal agents? Some observers had concluded that that anomaly alone was enough to suggest he was mentally unbalanced. The *New York Times* described his having "a long history of mental illness."

"So what was it all about, Gary?" I prodded when I finally sat down with him.

He responded by conceding that attempting to hijack an airliner was a pretty high-risk enterprise, so this inveterate risk-taker knew there was a good chance of failure. He also knew that most hijackings at that point were politically motivated. So in explaining his logic to me, he said something to the effect of "If I couldn't work my way out of this one, I knew I'd be doing some hard time. And I figured if the big black brothers thought I was a political prisoner, I'd be less likely to get my ass raped in the shower."

Notwithstanding the racism in this declaration, it is highly significant from a behavioral perspective. First, it shows that rather than being loony, Trapnell was fully rational and planning ahead for contingencies. So there goes the insanity defense.

It also helped us improve our hostage negotiation approaches and procedures. Regardless of the situation—airplane hijacking, bank robbery, even terrorist incident—if the hostage-taker makes a statement or demand that seems off the wall or out of character, the negotiating team has to think seriously about its actual meaning. Is the subject mentally decompensating due to stress or fatigue and therefore saying something nonsensical? Or is there a deeper meaning that can be used to defuse or end the situation without violence and bloodshed?

In this case, it showed that Trapnell understood he was unlikely to free himself from this scenario and was already considering the next step, which was his capture and imprisonment. This indicated the greater likelihood of settling the matter without hostages being harmed. In other words, at that point he was trying to mitigate the severity of the consequences of his actions rather than compound them. It also suggests to the negotiator that he or she may have something significant with which to bargain. Rather than concentrate on the ransom demand or where the hijacker wants the airplane to fly, a dialogue could be opened regarding why Trapnell was adamant that Angela Davis be freed. In this way, the negotiator could get down to his real concerns.

Similarly, at one point I went in by myself to interview Bruce Pierce, one of the assassination killers of controversial Denver radio personality Alan Berg. Pierce was a member of an anti-Semitic white supremacist group known as the Order whose members believed Jews were descended from Satan. As it turned out, Pierce agreed to the interview only so he could lecture and verbally abuse me and the FBI. Although this experience would seem to have been a failure, it was valuable in that I was able to gain insight into such a mindset—the demented focus and the dedication to a cause. So if law enforcement was in a standoff or hostage situation with someone like this, the negotiator's strategy would be to stall for time as he or she measured the gunman's dedication to his cause by restating or paraphrasing his content and prepare for a tactical response if negotiations were breaking down, before there was a loss of innocent lives.

WHAT WAS REALLY DOMINATING MCGOWAN'S THOUGHT PROCESSES AT THE TIME OF the offense? That is always the key question. I was hoping this would emerge from Joe McGowan when Andrew Consovoy spoke to him directly.

By this time, I had gotten to know Consovoy well enough to call him Andy, and I had come to feel a great respect for his intelligence, rigorous work ethic, and dedication to the hard job of seeing to it that the right people were kept in prison and the right people let out at the right time. He told me that he and other members of the parole board would be conducting their own interview in the next week and asked for my advice on how to go about it.

I said I thought he and his colleagues ought to keep up continuous questioning and let him keep talking. Eventually, there would come a point when the real fury underneath the calm exterior would surface. If they followed this plan, I thought there was a good chance that McGowan would get to the point where the real guy would come out and the parole board would gain additional insight and confirm my observations and recommendation.

When Wayne B. Williams was on trial in 1982 for the Atlanta Child Murders, Fulton County assistant district attorney Jack Mallard asked my advice on how to approach the defendant if his attorney put him on the witness stand and Mallard therefore had the opportunity to cross-examine him. I said first that I thought there was a good chance Williams might take the stand because I had detected in him a fair amount of intellectual pretension and superiority and a feeling that the criminal justice system was a bunch of bumbling Keystone Kops. He thought he could control the situation, even from the witness chair.

I suggested to Mallard that he come in close on Williams, violate his personal space, go through the case and his personal history in a *This Is Your Life* fashion, and sustain the tension by peppering him with questions until he had rattled Williams enough to catch him off guard.

When Williams did take the stand and Mallard got his chance, he did just what we had talked about. Finally, after several hours of sparring cross-examination, Mallard came right up to Williams, put his hand on Williams's arm, and in his low southern drawl, said, "What was it like, Wayne? What was it like when you wrapped your fingers around the victim's throat? Did you panic? Did you panic?"

In a quiet voice, Williams responded, "No." Then he realized what he had done and flew into a rage. He pointed a finger at me sitting in the courtroom and shouted, "You're trying your best to make me fit that FBI profile, and I'm not going to help you do it!" He started ranting about FBI "goons" and prosecution "fools." But it was the turning point in the trial. Several of the jurors who convicted him later said so themselves.

I thought the same tactic would work with Joe McGowan in the trial-like setting of the parole board hearing.

As I have said, I had a twofold aim in consulting on the McGowan case. The first was to help the New Jersey Parole Board in formulating a responsible and appropriate recommendation. The second was to learn as much as I could for my own work about the way this particular killer's mind worked. I was highly interested in whatever he had to say to the board after spending all those hours with me. I didn't hear right away, but eventually, after the decision was made, Consovoy related what had happened at his encounter in the Trenton prison.

The hearing had had one main purpose: to determine whether there was a "substantial likelihood that appellant would commit another crime if released on parole."

McGowan admitted that he had not been completely candid or forthcoming in his numerous therapy sessions. He acknowledged that in 1970—three years before Joan's murder—he had briefly dated a sixteen-year-old student. The girl never said anything to anyone, so he was never disciplined, though two other teachers had been fired for dating students. When he was asked why he would risk his career on so clear a violation of school policy, he said he now realized it was because he could be in a "superior position."

A former student who was at Tappan Zee while McGowan was there told Rosemarie about an uncomfortable encounter with him.

Although he wasn't her teacher, she needed him to sign a paper when she was mistakenly enrolled in a chemistry class as a freshman. This was about two weeks before Joan's murder, and because she felt intimidated by him, she brought a friend along. "The way he looked at me," she related to Rosemarie, "I felt like he was a giant who wanted to eat me up!" This is kind of vague, but the fact that this young woman felt threatened is clear.

Right from the beginning of the interview, Consovoy reported, he'd had the impression that McGowan considered himself intellectually superior to the board members, much like Wayne Williams. This didn't surprise me. What did surprise me was that in recounting the details of the murder, McGowan stuck to the excuse that he had not had the proper denominations of bills to pay Joan, so that was why he had asked her to accompany him to the lower level.

How could he stick to that version, I wondered, when I thought we had so completely undermined it during my interview? But the more I thought about it, the more it fit into place. This was a guy who was used to manipulating facts to serve his own view of things. And if he did feel intellectually superior to what he perceived to be ham-fisted, politically appointed board members, then never mind that he had already admitted to me that as soon as Joan came to the door, he knew he was going to kill her. He could tell them anything he wanted that enhanced his claim that the murder was a spontaneous outburst of madness.

Consovoy told McGowan point-blank that he did not believe the narrative of the murder as he had confessed to it. McGowan casually agreed that maybe it wasn't the entire or completely accurate story but wouldn't let on that he was at all bothered by being challenged.

Rather than pursue this line, Consovoy mentally stored it away for later use and switched to the topic of McGowan's upbringing. He got him to go through his early and formative years and to talk about his mother and father and their relationship with him.

Consovoy remembered an incident from the file in which McGowan's younger brother, only a toddler, had been seriously ill from a congenital condition. McGowan had mentioned in one interview that in the boy's final days in the hospital, his mother would not let him come upstairs to see his brother and made him wait in the lobby. Apparently, she thought it would be too traumatic for young Joe to see his brother so near death. And when he brought the incident up, Consovoy could see that McGowan was bothered by it, so he decided to pursue that line and see what happened.

Eventually they got to the moment I had been anticipating. We had focused on the time between when McGowan's engagement had broken off and the murder, with the other teachers going on the Easter trip without him. During that time—from Valentine's Day to Easter—people at the high school whom the parole board interviewed had noticed a change in his manner and behavior. One individual reported that he had "started acting really weird."

McGowan said that by the time of the murder, he had been thinking about committing suicide because he considered himself such a total failure. "I'm not dating anyone. I have no relationships. I'm not going anywhere. I mean, here it is Easter, and most of my friends are taking off on an Easter trip . . . down to Florida or Mexico, or going somewhere, and I'm sitting around doing nothing." He added that the reason he had not gone through with the suicide was that he was "too cowardly to do it."

The way he described it, "The doorbell rings, this poor little girl is standing there, and the thought flashes across my mind, 'Well, you can't kill yourself. Can you kill that?' "

Andy reacted with something to the effect of "Are you trying to tell me you killed a little girl over a trip?"

" 'Well . . . ' He started hemming and hawing," Consovoy recalled while recapping the scene. "Why don't we go back through this, then, because I don't believe it. I can't understand it. If everyone was going off in couples and you're going off single, and that bothered you so much, then what in the world was going on?"

Then it came out.

"He told me the whole story about planning on getting married," Consovoy related. "He met this girl. They fell in love, et cetera, et cetera. Then he brought her in and introduced her to his mother and grandmother.

"And then his mother said, 'You're not getting married.' He said, 'Yes I am.' And she said, 'Well, you can get married, but if you do, you can pack up and leave now, take your bride with you,' or whatever she called her. 'Don't ever see me, don't ever talk to me. You're cut out of the will. You have no money coming to you. Good luck!' " Or as Consovoy succinctly relayed it: "It's either her or me."

The fiancée had not dropped Joe. He had dropped her!

Consovoy continued, "He didn't say why his mother reacted that way; he just said that's how it was. And Joe said goodbye to the girl and stayed with Mama. Aside from the resentment McGowan must have felt toward his mother but was unable to express to her, he may also have felt that this was his last, best shot at real happiness."

With this breakthrough, Consovoy thought, *This is good. Let's keep after it.* He said, "Okay, now we're going to get serious about your mother. It seems to me every problem we've discussed since I've been sitting here has to do with you and your mother. It's almost like a love-hate thing.'

"And he shook his head and said, 'No, no, no, I never hated my mother!'

"I said, 'Face it: Every difficulty you had, everything that went wrong for you, it's somehow traced back to your mother. Your father died early, your brother . . . ' I looked at him and said, 'How long have you been in therapy?'

" 'Oh, I've been in therapy twenty years.'

" 'And what do you talk about at therapy?'

He told Consovoy he talked about why he did it and how he'd learned his lesson and why he'd never do it again, how he can avoid these things. So Consovoy asked him, "Do you ever talk about your mother?"

"And that's when he turned—right then and there. I'll never forget it as long as I live. I said, 'Do you ever talk about your mother?' and he looked at me with a face that was not 'nice Joe McGowan'—this was a very cold face—and said, 'My mother is off-limits!' It was almost threatening to me; that if I continued the discussion he'd leave. 'My mother is off-limits!'

"I said, 'Wait a minute! You're telling me that you've achieved everything you have to, that you're completely rehabilitated in keeping with the seriousness of the crime. And it was a highly serious crime, so you have a high standard of rehabilitation to meet, and you're telling me you're completely and fully rehabilitated after murdering and raping a little girl and putting her in a garbage bag and driving out of state, that you never discussed your mother?'

"He said, 'I said, My mother is off-limits.' "

But Consovoy persisted. Ultimately he got McGowan to concede that his "overwhelming feeling of sexual inadequacy" could be traced to his mother.

"I said, 'Let's go back.' We rehashed the mother's ultimatum. I said, 'How angry were you?'

" '*Very* angry.'

" 'Did you stay angry?'

" 'Yes. I was angry for two weeks. I was angry all over the place.'

" 'But you couldn't show it because you were afraid of your mother?'

"He said, 'Yes, I was,' and he was really getting mad then, because it was bringing up all his sore spots."

In our research, there is a strong correlation between domineering mothers and men who grow up to be predators. Though the vast majority of those with such mothers do not grow up to be offenders, of those who do, the domineering mother constitutes a significant influencing factor.

During the filming of *The Silence of the Lambs*, the FBI happily cooperated with the producers and even allowed scenes to be shot at Quantico. Despite the notoriety of Dr. Hannibal Lecter, the central crime in *Lambs* is perpetrated by Jame Gumb, known as Buffalo Bill and brilliantly portrayed by Ted Levine. Bill is a composite of three actual serial killers—Ed Gein, Ted Bundy, and Gary Heidnick, all of whom we studied in great detail at Quantico.

Director Jonathan Demme, with whom I developed a close relationship, asked me to coach Ted and explain to him what would be going through the mind of a felon like Jame Gumb/Buffalo Bill. As I told the New Jersey Parole Board in the McGowan case, my cardinal rule is that to understand an artist, you have to view his art. Likewise, with a predator, to understand him, you have to understand his "art," because that's what it is to him. The rest of his life wouldn't much matter to him; it would be tedious and unexciting.

In the case of Buffalo Bill, understanding the "art" was relatively straightforward, because Bill actually was creating something material: a woman's suit made of real women. This suggested to me, I said to Ted and Jonathan, that the root of his psychopathology went back to his mother, as it did with Ed Gein. By assuming a woman's skin, he would in his own mind be re-creating his mother's power over him in himself. He would have felt that life had not been fair to him and he was therefore justified in anything he did to others.

Though McGowan wasn't so literal, his anger was just as real.

After finally breaking through on the subject of his mother, Consovoy said, "So now, let's go back through the [murder] story. Let's be honest: Whoever came to that door was going to die.

"He said, 'That's right.'

" 'Did it matter who?'

" 'No, not unless it was an armed policeman.'

" 'Well, that's good thinking on your part. Are you sure of that?'

" 'Well, I think so.'

" 'But even then, you're now faced with this little girl that you've decided to kill; you're going to kill her. But you have to get her inside. She lived down the block; you couldn't go chasing her all over the lawn. What are you going to do?'

"And he looked at me with that same look again and said, 'You seem to forget that I'm a teacher, and I can command a child, particularly a child of that age, with my voice.'

"I said, 'Show me.'

"He goes, 'Joan, you'll have to come in the house,' or something like that—some teacher's voice. And anyone who's been in school recognizes the 'teacher's voice.' And she came right in. I said, 'Once that happened, how long until she was dead?'

" 'It was very fast.'

" 'So you had already made the decision.'

" 'Yeah.'

" 'What is the stuff with the change? You never even made change.'

"He said, 'No. No.'

" 'The rest was just details?'

" 'Yeah.' The whole story that he'd maintained for twenty-five years until John interviewed him just collapsed in a heap and he couldn't get it back, because everything else he said just fell through, and it made our own work a lot easier.

" 'So, what about the rape?' He said the rape was just opportunistic. The rape was just because she was a seven-year-old girl. He wasn't going to rape a man in a similar situation; that was not the deal. The rape was just another manifestation of anger. When he opened the door, she was a dead person. He was just so angry, I think that his whole world fell in on him—all the things that had happened all of his life. I mean, how many people at his age don't get married because of their mother? He wasn't a kid. Something set him off. I think that trip did have something to do with it. He was all by himself. He was acting strangely. The trip represented the fact that he was a failure, and maybe a coward, because he couldn't stand up to his own freaking mother, for god's sake.

"And at that point, we just got back into the more mundane details of the crime. I just wanted to clear up the record on all of the details, like burying the body. The way he disposed of the body, he definitely planned to get away with it. As far as all of the other things, like the confession, I thought the police reports spoke for themselves."

Had McGowan not been apprehended after murdering Joan, would he eventually have worked his way up to killing his mother, as Ed Kemper did? Probably not. His mother and her treatment of him wasn't the all-consuming emotional displacement for him that it was for Kemper, and he seemed to have a lot more conflicted feelings. But that doesn't mean the resentment and anger would ever have gone away, as evidenced by his reaction to Consovoy's probing.

The hearing lasted most of the day. When it was over, the board members met and reviewed their findings.

On November 6, 1998, the New Jersey Parole Board issued its official report denying parole to Joseph McGowan.

In the statement explaining its reasoning, the board cited several factors. First, there was the brutality of the crime itself. Second, it found that McGowan's lack of insight and concern over what caused him to commit the murder was "extremely disconcerting." The board felt he had made little progress in addressing the issues that led to the crime, and this was largely due to his lack of candor and honesty with the various psychiatrists, psychologists, therapists, and other authorities with whom he had spoken over his years in captivity. The board noted that while McGowan had taken part in many hours of therapy, no more than four had focused on his anger at his mother, which he had finally conceded, as the report stated, was a "primary motivator behind the commission" of the murder.

The key consideration, however, was the board's opinion that the state of mental and emotional health was not "far different than it has been in the past, and it remains substantially likely [McGowan] would commit a crime if released on parole."

The other issue to be dealt with was the future eligibility term, or FET—known informally within the prison system as a "hit." The board referred that determination to a three-member panel of board members—a standard practice.

On January 7, 1999, that panel imposed a thirty-year FET, theoretically meaning the prisoner would not be eligible for parole again for three decades. In practice, that was not exactly the case, since the state appeals court had directed the parole board to review its initial 1993 decision, which meant the hit meter started running from that date. Also, like all parole-eligible prisoners, McGowan received statutory credits for work and good behavior. Even more important, as a safeguard against abuse of the system, offenders in McGowan's category were legally entitled to an annual review hearing, at which point the parole board could institute a new evaluation if it believed his situation had changed.

This may all seem like an excursion into the bureaucratic weeds, but it is this kind of procedural process that determines freedom or continued incarceration. And by extension, it therefore determines whether the public is subject to the risk of individuals who have shown a previous propensity toward violence.

McGowan appealed the FET to the full board, which affirmed the three-person panel's decision on August 2, 1999.

He then took his complaint to the Appellate Division of the Superior Court of New Jersey. In the *McGowan v. New Jersey State Parole Board* ruling, his attorney argued that he had been a model prisoner for almost thirty years, there was no evidence he would reoffend if released, and therefore the board had acted in an arbitrary and capricious fashion. Suffering through a particularly bad period of her chronic condition, Rosemarie issued a victim impact statement from her living room.

On February 15, 2002, the court released its ruling, which upheld the parole board's denial and declared, "The decision to impose a thirty-year FET is within the Board's discretion and is supported by substantial evidence."

9

JOAN'S LEGACY

We'd achieved our objective: Joseph McGowan would remain safely behind bars for the foreseeable future. And in 2009, with new members, McGowan received a thirty-year future eligibility term and this time did not bother appealing the parole board's latest ruling, setting his next eligibility date as August 2025 and bolstering the likelihood that he would remain there for the rest of his life.

As part of the movement Rosemarie led, an 80,000-signature petition and 300 letters were sent to the parole board. When she got the call informing her of the board's decision, she draped Joan's green poncho over her shoulders, the one Joan was wearing when she went out to sell Girl Scout cookies with her sister Marie.

"He's never going to get out, and this means we can stop having to fight this every few years, and it means justice for Joan," Rosemarie said. But this did not signal an end to the vision and the purpose she had set for herself.

In 1998, the twenty-fifth anniversary of Joan's death, and at the same time the parole review was under way, Rosemarie formally established the nonprofit Joan Angela D'Alessandro Memorial Foundation. Its mission was and is to promote child safety and protection, advance victims' rights, and help homeless and neglected youth. Her sons Michael and John help her administer it.

Supported by "Team Joan" volunteers, the foundation's Fun, Education, and Safety Program has helped underprivileged of the Father English Community Center in Paterson and Passaic, the YCS (Youth Consultation Service) Holley Center in Hackensack, Tails of Hope Foundation in Pine Bush, New York, and Hearts and Crafts in Hillsdale. The program gave them opportunities for fun and educational excursions every year since 2001, including New York City, Washington, D.C., Great Adventure theme park and the Jersey Shore as well as assisting victims and advocating for legislation. The program has been expanded to support Covenant House in Elizabeth, which helps young people ages eighteen to twenty-one secure more stable futures. Since 2016, Joan's Joy has been providing a child safety program for local schools that trains teachers and parents how to spot, report, and prevent child abuse.

Among the foundation's recent accomplishments was spearheading with local legislators the effort to have the New Jersey General Assembly introduce a bill that amends Joan's Law, raising the threshold age of a victim killed in the commission of a sexual assault to under eighteen for a life sentence without the possibility of parole. A professor from the St. John's University School of Law in New York has used a video interview with Rosemarie to teach her students about effective advocacy.

On the civil side, Rosemarie proposed and advocated the Justice for Victims Law, which the state legislature passed on November 17, 2000. It was signed into law in the Hillsdale borough hall. Rosemarie was too weak to attend but was represented by Michael and John. The new law eliminated the statute of limitations for wrongful death actions brought in murder and manslaughter cases, allowing victim survivors to sue convicted killers for inheritance or any other assets they acquire any time after the crime.

The following year, Rosemarie sued McGowan and won a $750,000 judgment. He did not contest the suit, though by that time, almost all of the funds he received from his mother's and grandmother's estates had been disbursed to a relative or spent on attorney fees. He is required to pay her every month out of his prison earnings, averaging about $14, and Rosemarie has assigned every penny to the foundation. Unfortunately, I didn't see that McGowan cared anything about Joan's loss of life or Rosemarie's feelings. If he cared about anything, it was the bother of being reminded and not having the $14 to spend in the prison canteen. What really bothers him is that he got caught and had to face the consequences.

Rosemarie's fight for justice started to get national attention. In 2004 she received an award from the Department of Justice Office for Victims of Crime for demonstrating extraordinary courage and heroism. John went to Washington on her behalf to accept it from Attorney General John Ashcroft because she didn't have the physical strength to go.

Still concerned that what had happened to Joan could happen to other children, Rosemarie persevered to bring awareness to this case that had focused society's attention on child safety. She continued in her quest to communicate her safety concerns to the Girl Scouts administration, and in October 2014 met with the chief executive officer of the Girl Scouts of Northern New Jersey and the chief girl experience officer from the national office. Her agenda included a request to end the practice of having the girls go door to door to sell cookies or collect money, citing a statistic she had been told about from the U.S. Bureau of Justice Statistics that fourteen-year-old girls are among the most vulnerable children to sexual assault. This ban on door-to-door selling is something Mark and I have advocated for a long time.

I didn't feel my job was over, either. If there were something more I could learn about a guy like Joseph McGowan and how his mind worked, that would always be valuable to me and the victims I endeavor to serve.

That opportunity arose in fall 2013, when the Joan Angela D'Alessandro Memorial Foundation presented a Celebration of Joan's Life and Legacy on September 7—what would have been her forty-eighth birthday—in a town near Hillsdale. It was a dinner dance and benefit for the foundation, and I was asked to give the keynote address. Mark and his wife, Carolyn, accompanied me, and we met and made contact with as many people as we could who had played a role in Joan's story.

The celebration was splendid. Michael was the host and MC and John the videographer. It was both heartwarming and thrilling to see firsthand the outpouring of love and admiration for Rosemarie and all that she had done since the movement began in 1993. We all wore green foundation wristbands and green ribbons in honor of Joan.

We spent most of the following day with Rosemarie, John, and Michael at her house. From the living room where she sat with us, Rosemarie can see down to the end of Florence Street and the house where her daughter was murdered. She was tired from all of the arrangements and festivities of the past several days—myasthenia gravis extracts a quick payment for any prolonged expenditure of energy—but she wanted to talk to us and go over the entire story.

She showed us Joan's room and unwrapped her carefully preserved Brownie uniform. Her tiny ballet slippers are preserved in the hallway just outside the kitchen. There was no way for us not to get misty-eyed. It was almost like being in the presence of sacred relics. On the wall of the dining room we saw framed the signed copies of the bills establishing Joan's Laws in New Jersey, New York, and nationally, signed by Christine Todd Whitman, George Pataki, and Bill Clinton, respectively. Then we retraced the steps Joan had taken between her house and McGowan's. It was chilling actually to see how close the two homes are.

One of the things that used to surprise me but no longer does when I am talking to the families—especially the parents—of murdered children was how much they often want to know about what was done to their child. Like most police officers, I tried very hard to spare them the horrible details. But in so many cases they wanted to know, as if to share the suffering and somehow take it upon themselves. I remember that Katie Souza insisted the funeral home allow her to view her eight-year-old daughter Destiny's naked body after the adorable little girl had been beaten to death by her aunt's boyfriend, so she could experience every individual wound Destiny suffered and recall it for all time. I remember Jack and Trudy Collins describing how they stared at the body of their twenty-year-old daughter Suzanne, a lance corporal in the Marine Corps, whose stunning beauty had been so shattered by her torturer-killer that before her burial at Arlington National Cemetery, her wake required a closed casket. And I remember them, years later, asking Mark for all of the details of the medical examiner and police reports he'd researched on her murder so that they could share her agony. In fact, the strongly religious Jack used to go without Novocain for dental procedures, asking God to retrospectively and proportionately alleviate some of Suzanne's final suffering.

So it was with Rosemarie. She wanted to know from us everything we'd learned in the case files about what Joan went through to add to her accumulated store of detail that strengthened even more her ongoing communion with her daughter.

She was particularly interested to know whether Joan fought and struggled with McGowan or submitted meekly. We told her that from Dr. Zugibe's medical report and from what McGowan had told both Andy Consovoy and me, it was clear that once Joan understood what was happening, she had put up a fierce resistance even though she was no match for the six-foot-two adult male.

Rosemarie wasn't really surprised that she fought for her life so bravely because Joan was never afraid to stand up for herself or others. "A schoolmate of hers spoke to me on the phone about fifteen years ago; it was the first time I had talked to her since Joan had died and now this young lady was all grown up. She told me that Joan would bring her into her group on the playground whenever she saw that she was alone and not playing. She said Joan made her feel accepted.

"She is my inspiration that is beyond words to describe," says Rosemarie. "And that's why I chose to fight for Joan's justice and to protect other children. It can't be done by staying quiet. It can't be done by thinking that somebody else is going to do it. You just have to do it."

On April 19, 2013, a second vigil was held at Veterans Park, this time to mark forty years since Joan's death. At this event plans were unveiled for a sculpture and garden that would share the significance of Joan's legacy with generations to come.

ON APRIL 3, 2014—THE ANNIVERSARY OF THE SIGNING OF THE FIRST JOAN'S LAW— near the town train station, Hillsdale unveiled and dedicated the stone sculpture and garden in memory of Joan.

The sculpture and garden were funded by Joan's Joy supporters. Many local businesses also pitched in, donating their time and resources. On the brick path leading to the sculpture is a custom-made park bench in green, Joan's favorite color. There is a white butterfly in the center of the back, and above it, a reproduction of Joan's signature from a butterfly she drew when she was four and a half. It is flanked by orange flowers taken from one of her drawings.

Speaking of Rosemarie, County Executive Kathleen A. Donovan said, "She has turned her grief into something that has been a real shining star and shining example for us to follow."

The side of the monument facing the street highlights a white butterfly and the legend "Remember Joan Today So Tomorrow's Children Will Be Safe." The side facing the station, where it will be seen by all visitors and residents returning home, bears the smiling, achingly evocative picture of Joan in her Brownie uniform and an account of her story. The Joan Angela D'Alessandro Memorial Foundation raised the funds for the sculpture. A beautiful and colorful garden surrounding the sculpture was planted and dedicated on June 27, 2014. And on

April 19, 2018, the Child Safety Forever Fountain was unveiled in the garden. Its continuous flow of water symbolizes the never-ending importance of child safety.

The carved-in-stone account briefly relates the story of how Joan came to be represented by a white butterfly. On a cold day in April 2006, during a visit to the site where Joan's body was found in Harriman State Park in New York, Rosemarie spotted a white butterfly hovering behind the split boulder where she was found. Having already invested great significance and spiritual meaning in the fact that Joan was murdered on Holy Thursday and discovered on Easter Sunday, Rosemarie took the beautiful creature as a sign that Joan's soul was happy. As she related the story over the following months and years, the butterfly came to symbolize Joan's unquenchable energy and spirit.

"What we're doing today, coming all together, is about social justice," Rosemarie said at the dedication. "When you see this sculpture, something is going to happen to each of you and I hope that you share it with people who aren't here."

While all of the others were praising her, Rosemarie was praising her daughter. "My main inspirational person is Joan. I would not have been doing what I'm doing to this day if it wasn't for her inspiration, and it's like she's telling me to do what I'm doing."

Despite this heartbreaking inspiration, Rosemarie wants visitors to think of the sculpture and garden "not as a place of sadness, but as a place of joy, of peace, of education, of awareness of child safety, and last, but not least, as a place of hope for society." It is also the site of the annual child safety fest fundraiser, furthering the foundation's mission.

Reading the words on the stone made me recall something—a small detail—from the day that Mark, Carolyn, and I went to Rosemarie's house.

We had completed our discussion, and Rosemarie, Michael, and John were serving us lunch on the screened porch at the back of the house. It was a cool fall day. Suddenly, from nowhere, it seemed, a pure white butterfly appeared and hovered in the air above us.

We all marveled at the "coincidence."

"See," said Rosemarie, "she's with us."

This many years later, Joan's influence is still being felt, not only on a public policy level, but on a personal level. Recently, a middle-aged man told Rosemarie how much he had loved being around Joan and playing with her.

He recounted that when he and his friends would get in a fight, Joan would break it up with the directive "Oh, come on. Let's play! Let's *do* something!" And that would take care of it.

The enchanting seven-year-old's words echo in Rosemarie's mind: *Oh, come on. Let's play!*

Let's DO something!

"And that's why I keep on doing."

"KILLING FOR ME WAS JUST LIKE SECOND NATURE"

10

ALL IN THE FAMILY

In real life, there are no violent criminals as brilliant or "glamorous" as Hannibal Lecter—anyone who says there are hasn't met them. Ever since I first started talking to killers, I've been determined to see and show these guys (and very occasionally, women) for what they really are. Such was the case when I was approached about doing an interview with an inmate named Joseph Kondro—not by a parole board but as part of a documentary television program.

A television producer who had followed my career approached me on behalf of MSNBC. He had been intrigued with the prison interviews with killers that my bureau colleagues and I had done, and he realized that this foundation of the behavioral profiling program—the one-on-one confrontation with a killer—could be turned into gripping television. As disgusted as I was with the spate of so-called reality programs that focused on manufactured adventure, phony romance, instant stardom, and above all, the systematic humiliation of seemingly ordinary people, I had to agree. The many times when I've faced off against killers who would have been just as happy to murder me as talk to me have been among the most intense experiences of my life.

Because let's be honest: The fascination with "true crime" is actually fascination with what writers and philosophers call the human condition. We all want to know and understand the basis of human behavior and motivation, why we do the things we do. And with crime, we are seeing the human condition writ large and at the extremes, both for the perpetrator and for the victim. In a very real sense, the television audience was after the same thing I was: a wider and deeper understanding of the criminal mind. And I do think there is a strong value in letting a large audience see what the face of evil really looks like. If we could find the right subjects to interview, what I wanted and what the television producers wanted would not be in conflict.

With the prison interview as the centerpiece of each program, the rest of the show would be supported with news clips, photos, and other documentary evidence of the crime and the killer, together with on-camera interviews with survivors, detectives, prosecutors, and others involved with the crime—similar to what an actual case investigation would be like. I agreed to do an episode, with the provision that I would have some level of control over the finished product. While I have no objection to playing off the ongoing fascination with the evil of violent crime as long as it leads to greater understanding and insight, I adamantly refuse to have anything to do with sensationalizing or glorifying the perpetrators.

The biggest problem with the television concept was a practical one: It's a lot harder to access serial killers for interviews than it used to be. Even for interviews that are strictly law enforcement related, the days of simply showing up at a prison and presenting your credentials as Bob Ressler and I used to do are long gone. Not only does the inmate have to give informed consent, but there are so many rules relating to safety, criminal process, and correctional system bureaucracy that getting in to see violent offenders is extremely difficult.

Because I was no longer in the FBI, I couldn't compel an inmate to talk to me, so we went through the whole procedure that began with writing to wardens and asking for their cooperation. This can be a big hurdle, because for obvious reasons, prisons are highly controlled and regimented. Everyone gets up at the same time, eats at the same time, and goes to bed at the same time, so conducting an extensive interview with one of the inmates disrupts the order of things.

I was looking for someone who fit the definition of a violent predator, but I was also trying to find an individual whose M.O. was different from anything I had encountered before, because I'm always trying for new insights that can expand our understanding of the criminal mind. And it was through this process of searching for someone willing to talk that I came across Joseph Kondro.

You never know for sure why certain incarcerated predatory offenders agree to talk to you. Some are bored. Some think you might be able to help get them out, like Joseph McGowan. There is always a flicker of hope, even for those sentenced to life without parole. Some think they can get better treatment or respect from the administration and staff by cooperating with federal agents or a former federal agent like me. Others relish the idea of reliving their crimes and feel it gives them status. And still others think of the interview as their self-analysis: They want an interpretation of the crimes they perpetrated. Some violent offenders already know the *why* of their behavior and embrace the challenge of whether I can discover their motive.

We knew that Kondro had turned down a number of interview requests in the past, which I suspected had something to do with unsolved crimes he didn't want to talk about. In any event, he would be spending the rest of his life in prison, but that life-span could be reduced considerably if he were charged with and found guilty of another murder that carried the death penalty.

I think the reason he agreed to speak with me was that my background was explained to him and he was "sold" on the insight he could provide law enforcement into his type of predator, which would help identify and catch other similar predators. I'm not sure he cared that much to learn about himself or even to help put others like him behind bars. But any prisoner with a reputation as a child-killer is not a popular guy, with either the staff or the general population, so he might have been doing some image enhancement by sitting down for the interview.

I didn't have access to the subject's prison file as I did when I was conducting officially sanctioned interviews, but I was provided with voluminous case materials as well as all of the media reporting, which I spread out on my dining room table. By the time I flew to Washington State, I felt I was pretty well on top of things.

Joseph Robert Kondro was serving a fifty-five-year term at the Washington State Penitentiary at Walla Walla. The former millworker, house painter, and laborer had avoided a death-penalty-eligible trial by copping a plea to the rape and murder of a twelve-year-old girl in 1996 and admitting to the unsolved murder of an eight-year-old girl in 1985.

What did these two victims have in common other than their preadolescent ages? Kondro was a close friend of both girls' families. And in my mind, that made him a standout interview subject. Over the course of my career, I have gotten used to gory crime scenes. It's far more horrifying to work my way into the psyches of those who produce them.

What kind of guy rapes and kills the children of people he knows well and who consider him a friend? What goes through his mind as he is planning and carrying out the crime? This is what I had to find out.

Kondro was also a prime suspect in the 1982 strangulation murder of eight-year-old Chila Silvernails in Kalama, Washington. Chila was last seen on her way to catch the school bus. Her strangled, naked body was discovered the next day in a creek bed. No arrests were made. Kondro had dated Chila's mother.

I had read an article in the *Seattle Post-Intelligencer* in which he claimed that since his incarceration he had returned to the belief of his Chippewa ancestors that requires people to atone for their misdeeds and try to correct them before they die; otherwise, their souls would be doomed to torment in the spirit world. I didn't know how much of this to believe, though criminals sometimes do find spiritual awakening in prison. But I intended to ask him why he had agreed to the interview. In the meantime, I had to prepare myself with everything I could find out about Joseph Robert Kondro and his crimes. Despite the fact that this interview was for a television program rather than a criminological study, I would conduct it the same way, following a huge amount of research, case file review, and preparation. I was talking to a killer, and as such I had to be prepared for whatever he turned out to be.

JOSEPH KONDRO WAS BORN MAY 19, 1959, IN MARQUETTE, MICHIGAN, TO A NATIVE American mother from the Chippewa tribe who already had six children and didn't feel she could care for another. She gave him up at birth and he was adopted by John and Eleanor Kondro, a white couple in Iron River, Michigan, where he grew up, before moving to Castle Rock, Washington. John was an aluminum worker for Reynolds Metals. Kondro later said his parents considered adopting him a mistake.

As a child, he had a difficult time adjusting, was fond of carrying a knife with him, and hung out with a group that tortured and killed small animals and pets they found in the neighborhood. Along with fire-starting and late-age bedwetting, this is one of the predictors of violent criminal activity that we see over and over again. Of the components of this "homicidal triad," cruelty to animals is by far the most serious.

The Kondros had tried to raise their son with a strict, middle-class upbringing, but he kept getting into trouble. His father had to bail him out of jail several times and paid for two stints in drug rehabilitation facilities.

By the time he reached his early teens, Kondro was molesting girls at school and in the neighborhood. What would become significant the more I learned was that as he grew older, his victim of preference remained the same age. He was accused of molesting girls and young women several times over the years, but most of the charges were not prosecuted.

On the afternoon of May 15, 1985, just before Kondro's twenty-sixth birthday, eight-year-old Rima Danette Traxler was on her way home from St. Helens Elementary School in Longview, Washington, a city of about 35,000 on the Columbia River in Cowlitz County. About two blocks from her house, Rima stopped to show a neighbor an art project she had made at school. The third grader was about four-feet-three and forty-five pounds—a beautiful blue-eyed girl with blond hair and a friendly disposition. She was wearing a pink shirt, a tan plaid skirt over white tights, dark brown clogs, and a belted knee-length coat. I stress the details of her appearance because this was the last time she was seen, alive or dead.

When her mother, Danelle Kinne, began to worry about her daughter's not coming home, she walked to the school to retrace Rima's route, but saw nothing. When she returned home, she called Joe Kondro, who was an old high school buddy of her husband, Rusty Traxler—Rima's stepfather—and a good friend of the family. Many years later, Danelle recalled that earlier on the day her daughter disappeared, Joe and Rusty had been sitting on her front porch, drinking beer and laughing while Rima sweated through mowing the lawn. They were making fun of her for being so diligent about keeping the yard neat.

After Danelle's call, Kondro came over and Danelle even used Kondro's cell phone to call police. As soon as the announcement of her missing child was published, police and community members launched an intensive search, similar to the one that had been mounted for Joan D'Alessandro. But they couldn't find any trace of her.

Around the time of the disappearance Kondro was seen in the vicinity, driving to a convenience store to buy beer and cigarettes. He was questioned by the police, but there was nothing to tie him to the missing child. The case remained open and unsolved.

Years passed and Kondro remained free.

Over a decade later, on November 21, 1996, also in Longview, Washington, twelve-year-old Kara Patricia Rudd and Yolanda Jean Patterson decided to skip out of their classes at Monticello Middle School. At the time, Kara and Yolanda were living in the same house together, along with Kara's mom, Janet Lapray, and her live-in fiancé, Larry "Butch" Holden. Yolanda was Larry's niece and he had guardianship of her and her brother, Nicholas. And until about a month before their house had another resident: Joseph Kondro.

Kondro was a close friend of Kara's mom and often stayed with the family. By this time, the thirty-seven-year-old Kondro was the father of six children by three different women, and he didn't regularly support any of them. Kondro had recently renewed his relationship—and periodic residence—with Larry and Janet because he was between girlfriends; indeed, he had become a regular enough fixture at the house that Kara called him "Uncle Joe." But his most recent stay had come to an abrupt halt after Janet and Larry threw him out when his drinking and drug use became intolerable. Janet later said Joe used to come on to her while Larry was away.

On the morning in question, Larry had dropped both girls off at school at 7:15. Around 7:30, a gold 1982 Pontiac Firebird pulled up to the sidewalk abutting the school parking lot; the car belonged to Kondro. According to Yolanda, when the girls spotted the car, she went over and leaned in the window on the driver's side while Kara got into the car on the passenger side. Shortly thereafter, Kondro rolled up his window, apparently so that he and Kara could have a private conversation. When Kara emerged, she told Yolanda she had asked Joe if he would take her out to pig farmer Pete's place in nearby Willow Grove so she could play with the piglets. She asked Yolanda if she wanted to come along, but Yolanda said she was afraid of getting in trouble with Larry or Kara's mom, so she declined and said she was going back to class. With that, Kondro's Firebird pulled out of the school parking lot. The last time Yolanda saw Kara, she was walking east on Hemlock Street, presumably to meet up with Kondro. Yolanda then went into the school building.

Like Rima eleven years earlier, Kara never returned home. Before that even became an issue, though, the vigilant school principal had called Kara's mother, Janet, to say that she was absent. When Kara did not return at the end of the day, Janet immediately thought of Kondro and even accused him of abducting her daughter—a conversation that was recorded by a malfunctioning answering machine in Lapray's home. For some reason, it kept running even after she picked up the phone.

Police instituted a community-wide search and released Kara's photograph to the Longview *Daily News*. Pete Vangrinsven, owner of the pig farm Kara wanted to visit, said he had not been home on November 21, but nothing looked disturbed to him and he invited detectives to look around. They found no sign of her. In the meantime, like Kara's mom, law enforcement focused on Kondro.

Joe Kondro was questioned by police and admitted he had spotted Kara and Yolanda outside the school that morning and had pulled up to talk to them. He agreed that Kara had asked him to take her to the pig farm but said he had refused, told her to get out of the car, and warned them to go back to school. He said that both young teens were good girls, but that every teenager gets into mischief. According to Kondro's account, he stopped at the Hemlock Store for a cup of coffee, and then drove out to Marthaller's Log Yard to look for a job. The office was locked, and though he did see men working out in the yard, the ground was muddy, and he didn't want to get out of his car, which nonetheless became stuck in the mud.

As part of their thorough examination of anyone connected with Kondro, Longview police department detectives interviewed Julie West, Joe's ex-wife, with whom he had two children and with whom he was currently living. He felt free to come and go because she had previously kicked her current husband out of the house. Julie told them Kondro was subject to violent rages and had assaulted her on a number of occasions, including tearing her clothes off one time when she was pregnant with their child, and ripping a sink off the bathroom wall. She finally had to obtain a restraining order against him, which led him to seek a divorce. That didn't end their relationship, though, and she got pregnant by him again one night when they had both been drinking heavily. Another time when he was at her house, he became belligerent and she threatened to call the cops. Kondro warned her that if she tried to, he would rip the phone out of the wall.

Julie said that at around 11:45 on the morning of Kara's disappearance, Kondro came over to her house to take their son to school. When Kondro returned about 12:30, he asked her to take a ride with him to apply for a job at Industrial Paints. As they passed Marthaller's Log Yard he commented that he had stopped there earlier to ask about employment but didn't get out of the car because of the mud. West thought this was curious, since she didn't notice any mud on the tires, fenders, or body of the Firebird.

Then there was the hairbrush. As she had climbed into the car, the passenger seat was all the way back, so she slid it forward. When she did, she noticed a hairbrush under the seat. She described it in detail: black with white bristles that were black at the tips. Some of the bristles were missing and others appeared chewed up. Later in the day, she spoke with Janet and asked if Kara had a hairbrush like that. Janet said that Kara always carried a hairbrush and that she thought it sounded like hers.

While Julie West was quite forthcoming, Kondro's current girlfriend, Peggy Dilts, was not. Together Kondro and Peggy had a daughter, Courtney, and Peggy did not want to cooperate with the police and forbade them from speaking to Courtney alone.

Peggy's lack of cooperation couldn't prevent other parts of Kondro's story from falling apart. Two clerks who worked at the Hemlock Store and knew Kondro said they had not seen him at the time he said he'd been in there. Similarly, an employee of Marthaller's said he would have seen anyone who came in the single entrance to the log yard, and Joe Kondro's gold Firebird had not been there.

Though Rima's disappearance from 1985 didn't immediately come to mind because it had happened so long ago, the idea that two pretty blond schoolgirls in the same town had disappeared under similar circumstances seemed like too much for coincidence. I suspect it had to be on some of the senior police officers' minds.

Ray Hartley, the lead detective on the case, learned that Kondro had been accused of molesting a friend's daughter two years before but had been acquitted and that he had been fired from a job in a lumber mill for doing drugs in the parking lot. Around the same time, he had trashed and wrecked the inside of the home of a woman he was dating and threw her pet cages out into the yard. And here's the part I find incomprehensible and so troubling every time I hear something like this: After the incident, the woman continued seeing Kondro.

Another woman, Crystal Smith, who had dated Kondro the previous spring and had also given him free run of her house, told detectives that he got mean when he drank and even referred to himself as Diablo at such times. She recalled a barbecue in the summer when he had had too much to drink and started slapping around another girlfriend, Vickie Karjola. Smith had to get in between them to stop Kondro's assault. Kondro was apparently so casual about his female relationships that when detectives questioned him and asked Crystal's last name, Kondro replied, "I don't know. We are just good friends."

While the police were assembling an overall portrait of Kondro's character as violent and dangerous, they continued to turn up more useful information about Kara's previous interactions with him. When a detective sergeant questioned Yolanda, she told him that she and Kara had actually skipped school a few weeks earlier so that Joe Kondro could take them out to an abandoned house near Willow Grove where there were a lot of cats and kittens. Kara had wanted to take one of the kittens to give her mother as a birthday present. On that day, he had told the girls to walk away from the school—much as

Kara had the day she went missing—so he could pick them up where no teachers could see them. When the detective asked if Yolanda had ever been in a car with Kondro before that, she said he had taken her, Kara, and his daughter Courtney swimming and camping along the Toutle River, up Interstate Highway 5. It had been cold, and they'd stayed only one night.

When the police were finally able to interview Courtney at the police station in the Hall of Justice, she admitted that her father could be "kinda mean," and that he had slapped her on occasion and thrown her around for talking back to him. He had hit her sister April in the head with his open hand a couple of weeks before. She also said that until he moved in about two months before, she hadn't really known him well, to the point that she called him Joe rather than Dad. Courtney said she overheard Kondro on the phone conversation he had with Kara's mom, Janet, in which she said she was going to call the cops on him. "Last time I saw her she was in my car and I told her to get out," Courtney reported her father as saying.

Like her daughter Courtney, Peggy Dilts finally had to speak to police at the station, and when she did, she revealed that Kondro had asked to borrow shovels from her garage. When police checked the garage, two shovels were, in fact, missing.

ONE OF THE THINGS THIS CASE SHOWS IS THAT NO MATTER WHAT SUSPICIONS POLICE or individual citizens may have—even if they have a pretty good narrative of the crime in mind—none of it means anything unless solid evidence can be developed. I often have readers and audience members at my lectures say whatever case I have been discussing didn't seem like a very difficult one to solve. And in one sense they are correct. Not every case involves complex profiling and investigative analysis. And certainly, not every case is mystery novel material.

But Mark and I have run experiments from time to time where we start by telling our audience who the offender turned out to be and then go through the case from the beginning. When we finish, most of our listeners think the case was pretty straightforward and don't see why the police had any trouble solving it.

Then we'll take the same case with a different audience and not reveal who the UNSUB turned out to be. Given the identical case narrative, those groups almost never come up with even the type of person who committed the crime, even if we give them a suspect list.

This is how it was with identifying the main suspect in the Atlanta Child Murders case, Wayne B. Williams. While his profile and apprehension may seem obvious in hindsight, at the time they were anything but.

The Atlanta Child Murders of 1979–1981 put us on the map with law enforcement agencies around the country and overseas. More than twenty African American children and adolescents—mostly male—had gone missing and turned up dead. Many in the police department, the media, and the community at large were convinced the murders were being perpetrated by a Ku Klux Klan–like hate group, intent on intimidating the southern city against its progressive views.

This was one of the ways we were able to get involved with the case, since it was a possible violation of federal civil rights laws. Also, since there were missing children, Attorney General Griffin Bell ordered the FBI to try to determine whether they had been kidnapped. Following the notorious kidnapping of aviation hero Charles Lindbergh's baby son in 1932, kidnapping was made a federal offense, allowing the FBI jurisdiction after twenty-four hours had passed. The Atlanta murders were given a bureau major case designation: ATKID.

But when Roy Hazelwood and I went down there at the request of the Atlanta police department and analyzed the cases, we became convinced of two things pretty quickly. First, these were not Klan-like murders; there was no symbolism, no behavior intended to intimidate or cause fear, and no signature or taking credit for the crimes. Moreover, when we visited the sites where the victims had been abducted and/or their bodies found, it became clear that any white person in these overwhelmingly black areas would have stood out and been witnessed by someone, since these areas tended to have activity around the clock. The individual or persons who abducted these young people caused no such sensation. We thought there was a good chance, therefore, that the murderer was African American, even though nearly all the serial killers we had studied thus far had been Caucasian.

In a workroom the police had dedicated to the murders, we went through every case file, read statements from witnesses in the areas where children disappeared and their bodies were found, studied crime scene photographs, and reviewed autopsy protocols. We interviewed family members to see if there was any common victimology.

Most of the victims were pretty streetwise but otherwise naive about the world outside their own neighborhoods and therefore susceptible to the right lure or come-on. Most were also living in pretty significant poverty, so it probably wouldn't take much to get them to go with a stranger for a modest enticement. To test this out, we had undercover police officers—both black and white—offer neighborhood kids five dollars to come do some job. It worked almost every time, plus proved to us that white men were noticed in these neighborhoods.

In our investigation we did think that two of the victims—both girls—were not part of the overall pattern, since the abduction style and victimology were different. In investigating multiple murders, you have to be extremely careful not to succumb to either linkage blindness or connecting cases that may not be related.

Even though the entire raft of unnatural child deaths was being attributed to one individual or group, we thought a number of the cases bore no evidentiary relationship to the main group. Some could have been copycats, and others simply unrelated child murders that happened to occur around the same time. Records indicated that every year there were about ten to twelve child homicides in the city. Most were personal cause classifications, in which the offender was related to or knew the victim.

We began to construct our profile. Though the vast majority of our serial killer cohort had been white, we'd also learned that these predators tended to hunt for victims within their own race. We were therefore pretty convinced we were dealing with an African American male, since female killers were so rare, and we thought a man would be able to exert more authority over these kids. He would be in his mid-to-late twenties, and homosexually attracted to the young boys, though the lack of sexual assault showed him to feel inadequate or ashamed of his own sexuality. Since the crimes took place at various times, we didn't expect him to have a steady job, or else he might be self-employed. We expected him to be of above-average intelligence, but an underachiever. And with the sense of authority he would have to exert over his victims, we thought he would be glib and maybe a law enforcement wannabe. If this were true, we would expect him to drive a large, police-type car and possibly have a large dog.

A break came when a tape recording, purportedly from the killer, arrived at the headquarters of the Conyers, Georgia, police department, about twenty miles from Atlanta. There was general excitement, but when I listened to the tape back at Quantico and heard a white man's voice, I was pretty sure it was a fake. But the speaker mentioned the latest victim and said his body could be found along a certain stretch of Sigman Road in Rockdale County. From the tone and psycholinguistic analysis, I thought this guy felt superior to the police. So I advised them to play to his belief and look on the *wrong side* of Sigman Road. If he was there watching, maybe they could catch him.

The press covered this search heavily, and as I had suspected, no body turned up. Sure enough, the guy called back to tell the cops how stupid they were. The "stupid" cops were prepared with a phone trap and trace and got the guy, an older white redneck, right in his house, getting rid of that nuisance. Just to make sure, they also went back to make sure there was no body on the *right side* of Sigman Road.

Soon after, though, the body of a fifteen-year-old black boy was found on Sigman Road, which told us something critical: the UNSUB was reacting to the media and trying to show his own superiority. With that in mind, we suggested several proactive ideas to the police, including hiring amateur "security guards" for a large concert benefiting the victims' families. But by the time my idea was approved by the assistant attorney general, it was too late.

When the next body was found, the medical examiner announced that hair and fiber matched that of five previous victims. Since we knew the UNSUB was following the media, I was convinced the next body would be dumped in a river, where such evidence would be washed away. It took a while to organize all the local law enforcement departments to a river surveillance. By then, a thirteen-year-old boy was found in the South River and then two more, a twenty-one-year-old and another thirteen-year-old, had been discovered in the Chattahoochee, the waterway that forms the northwest border between Atlanta and Cobb County. Unlike the previous fully clothed victims, these three had been stripped to their underwear, presumably to remove hair and fiber.

Over a month later, the local agencies were losing patience with the river surveillance when a police academy recruit named Bob Campbell, on his final shift on the Chattahoochee beneath the Jackson Parkway Bridge, saw a car drive across the bridge and stop briefly in the middle. Campbell heard a splash, directed his flashlight to the water's surface, and saw ripples. The car turned around and drove off, where a stakeout car was waiting that Campbell directed to follow.

The driver was a twenty-three-year-old African American named Wayne Bertram Williams, who politely told the officer he was a music promoter who lived with his parents. When the body of a previously missing twenty-seven-year-old black male was found downstream, Williams was placed under intense, "bumper-lock" surveillance.

Williams fit our profile extremely closely, including the police-type vehicle and the large dog. He considered himself superior to the authorities and handled his initial interrogation glibly. When police obtained a search warrant, they found hair and fibers in his car that matched those from the murders we had concluded were linked together.

Williams was tried and convicted for some of the Atlanta child murders. But here we come to the second point of which we were convinced:

In one case in which a teenage girl had been abducted and strangled with an electrical cord, we were sure the offender was a man with documentable mental illness, who had almost certainly spent time institutionalized. The police came up with a suspect who fit our profile, even to wearing the same type of electrical cord to hold his pants up rather than a belt. There was no way, however, to connect this man conclusively to the murder, so it was never brought to trial.

As the Atlanta Child Murders case was wrapping up and Roy Hazelwood and I were getting ready to leave the city, we were talking to the task force psychiatrist and I explained to him how we came to our conclusion about the then-unknown offender.

"How do you know?" the psychiatrist asked.

"From the way he committed his crime," Roy responded. "We try to think the way he thinks." We had learned this from Gary Trapnell.

This seemed to intrigue the doctor. He asked, if he were to administer a psychological test to us, would we be able to score as if we had any mental disorder he specified? We said we thought that we could.

He had us each go into a separate room and we had a go at the Minnesota Multiphasic Personality Inventory (MMPI)—the most widely used standardized psychological test for adults. We both scored as antisocial personality disorder–psychopathy, coupled with paranoid ideation. The psychiatrist professed surprise. Roy and I were both pretty proud of ourselves. We had proved we could think like the worst of 'em.

The results of this case might all seem clear-cut and obvious in retrospect, but in real time, it was anything but. We could suspect Wayne Williams all we wanted, but until the police could get actual evidence on him, they couldn't arrest him. Then the prosecution team still had to build the case against him. We may have helped to single out his profile, but that was of course just the start. Putting him behind bars required more than just psychology. And while suspecting someone might be okay for true crime TV pundits and online discussion groups, it doesn't hold any weight in the real-life criminal justice process. And as I was to learn, Joseph Kondro had thought through this notion quite coherently.

Like Wayne Williams, Kondro continued to be cooperative with the police when he was asked to come down to the station house for questioning. The only times he would become angry were when the detectives would catch him up in a seeming inconsistency. At the conclusion of the session, acting on instructions from the district attorney's office, Detective Jim Duscha, the lead detective in the case, warned Kondro not to have any contact with Julie West—not to go over to her house, not to speak to her in person or attempt to contact her by telephone. Kondro said he understood.

The next day Duscha received a call from Julie. She said that Kondro had called her that morning, asking what the police had said to her and what she had told them. He told her not to say anything more to the cops nor to tell them about this conversation, since he wasn't supposed to be talking to her. She responded that she would tell the police everything she knew since she had nothing to lie about.

She told Duscha she was very afraid of Kondro because of his violent nature. "I don't know what he will do when he finds out I talked to the police," Duscha quotes her as saying in his written report.

The detective responded to the situation right away and went to the home of a judge to secure an arrest warrant. That afternoon, he and Detective Sergeant Steven Rehame, the case supervisor, drove to Crystal Smith's house, where Kondro had last been seen. They knocked on the door and Kondro answered. They said they were there to arrest him for tampering with a witness. They handcuffed him and put him in the back seat of the patrol car, where they read him his Miranda rights.

In the interview room at the Hall of Justice, Kondro denied having spoken to Julie West or even knowing she was a witness. Finally, when he felt he had heard enough, Detective Duscha asked Kondro why he would not tell him the truth.

Kondro hung his head for a few seconds, then looked up and said, "I really need a lawyer." The conversation stopped and the detectives took him to the county jail and booked him on $25,000 bail. It was soon doubled.

11

THE ABANDONED VOLKSWAGEN

That same day, Crystal Smith gave a statement to Duscha in which she said Kondro had told her Janet Lapray and the police suspected he had kidnapped Kara. He repeated the story to her about seeing the two girls in front of the school, talking to them, telling Kara to get out of his car, then going to look for work.

Smith asked him what he was going to do if they charged him.

"They ain't got shit. I'm sticking to my story," she said he repeated more than once. She then related a time the previous summer when they had been camping at Battle Ground Lake and were playing a game in the woods. She asked Joe what he would do with a body.

"No body, no witnesses, no evidence," he replied.

Dog teams repeatedly searched areas that police felt Kondro might have taken Kara but came up with nothing.

By this time, police had located a witness who had gone to high school with Kondro and was present at the Oregon Way Tavern about six or seven years earlier when Rusty Traxler, Rima's stepfather, had yelled at Kondro, accusing him of killing Rima. Kondro had told him to shut up and a fistfight ensued. The more the police learned about

Kondro, the more violent incidents emerged. Elizabeth Ann Ford, another woman with whom he had lived on and off for about seven years and with whom he had a child, said she would kick him out of the house every time he would start drinking heavily. He had gotten into a fight with her brother and broke his jaw and three ribs. He had also torn a woodstove away from the wall and thrown it at her brother in an uncontrollable rage. She ultimately got the sheriff's office to tell him to leave her alone.

And the law was finally catching up with Joseph Kondro. In December 1996, less than a month after Kara's disappearance, he was arraigned in Washington State Superior Court in Clark County for the molestation of a seven-year-old girl and the rape of a ten-year-old girl, both in September 1991. The prosecution contended that while visiting a friend, Kondro had sexually molested the girls as they slept on the living room floor. Trial was scheduled for the following May.

MEANWHILE, THE POLICE IN LONGVIEW CONTINUED THE HUNT FOR KARA, REVISITING several locations that witnesses and informants had told them Kondro liked to frequent. One was a vacant, decaying house on Mount Solo, west of Longview. It was a place kids liked to play, and several members of Kara's family, including her uncle, had come upon the area during their three-week search following her disappearance, but had found nothing.

On January 4, 1997, less than two months after Kara's disappearance, police were searching a remote wooded hillside up Mount Solo Road from the house. They came upon a ravine and spotted an abandoned and rusted red Volkswagen with no tires or wheels, facing south, with old Washington State license plates still attached. Inside the car, the police searchers found Kara's black Reebok T-shirt and bra and then discovered the body of a female underneath the passenger side with her head toward the rear and her feet beneath the passenger-side door.

There was an impact mark on a tree directly opposite the top of the driver's side doorframe and a corresponding dent on the car. This suggested the vehicle might have been tipped onto the driver's side against the tree, so the body could be placed underneath, and then dropped onto the victim.

Detective Sergeant Rehame requested state forensic technicians come out and process the scene. When they arrived, Longview police officers used a winch to tip the car back up against the tree so they could get to the body. They then took all relevant samples before cutting down the tree to remove the Volkswagen from the immediate scene and process further.

The upper part of the torso was badly decomposed and several of the ribs showed evidence of animal predation. But the lower half of the body had been well preserved by being completely under the car. Underpants and a pair of black shorts matching the description of what Kara was wearing were on the body. After collecting samples in the immediate vicinity, the crime scene techs loosened the dirt under the corpse so that they could slip two body bags over it, one from each end. They were taped together and sealed, and the body transported to the state crime lab. Dental records confirmed the body as Kara's. Cowlitz County coroner Gary Greig declared that she had died from "homicidal violence by unknown means."

Pieces of physical evidence, including all of the clothing found on and around the body, were sent to an independent lab and the San Diego Police Department Forensic Biology Unit for analysis. Semen deposits on Kara's body and clothing tied Kondro directly to the murder.

On January 27, 1997, Cowlitz County prosecuting attorney James J. Stonier presented in Superior Court a document known as a criminal information charging Joseph Kondro with aggravated murder in the first degree in the death of Kara Rudd. Kondro was imprisoned in isolation for his own safety awaiting trial; even violent incarcerated criminals hate child molesters and killers.

As prosecutors prepared a death penalty case against Kondro, he remained in the Clark County jail in Vancouver, Washington, with his murder trial scheduled to begin in July 1998. In the meantime, after Cowlitz County prosecutor Sue Bauer consulted with Danelle Kinne, Rima's mother, they offered him a deal: admit to both the Kara Rudd and the Rima Traxler murders and tell investigators where he disposed of Rima's body and they would not seek a death sentence.

In May 1997, Kondro stood trial in Clark County on the molestation and rape charges. He was convicted by the jury on both counts in less than two and a half hours of deliberation and sentenced to a combined 302 months in prison.

After some thought, and, perhaps, seeing how credible he had been to one jury, Kondro took Sue Bauer's deal. He said it was not just the prospect of execution that motivated him. He was also concerned that his children not have to testify against him and that the families of the two murdered girls should have closure.

Whenever I hear a killer say he has done something for anything other than his own self-interest, I am extremely skeptical.

IN MORE THAN TWENTY HOURS OF INTERVIEWS WITH LONGVIEW POLICE DETECTIVE Scott McDaniel, Kondro compared himself to an alligator that stays on the bottom of the pond until he gets hungry. Then he rises to the surface. It was an image that was hard to shake as I read his account of what happened to Rima all those years earlier.

Any murder of a child is horrifying and devastating, but the most disturbing detail of Rima's murder was how Kondro orchestrated it. This particularly significant—and despicable—element of the crime was revealed by Kondro's confession: that is, how he got the eight-year-old to go along with him. It turned out that Rima's mom had given her a secret password in case anyone approached her and tried to get her to go with him. The password was *unicorn*, and if the other person didn't know it, she would know not to trust him. Joe Kondro was such a good friend of Rima's parents that her stepdad Rusty had told him the password.

He told Rima he'd been sent by her parents to take her swimming and that they would join her later. When he saw her walking home, he later said, "I pulled over and thought, *If she gets into my car, I'm going to take her out to the woods.* And she just jumped in."

Kondro also told investigators that the day he had taken Kara and Yolanda out to the abandoned house was a "test run."

"I was planning on raping and killing them both," he said to law enforcement.

He had decided on the dump site for the body in advance. He confirmed that he beat Kara severely before raping and strangling her. Remember, this is a girl with whom and with whose family he had a close personal relationship—he was "Uncle Joe" to her. Yet he thought nothing of beating her viciously, strictly as a means to a perverse end before he sexually assaulted and murdered her. Though he was by no means insane, this is about as depraved as anything most of us can conceive. Then he went to Peggy Dilts's house, showered and scrubbed himself clean, washed his clothes, and threw away his shoes.

On February 26, 1999, Kondro entered his plea of guilty before Judge Jim Warme to first-degree felony murder in the death of Kara Rudd and second-degree intentional murder in the death of Rima Traxler. In the presence of his victims' relatives and friends, Kondro read his confession in open court. On March 5, Judge Warme sentenced him to fifty-five years. That sentence would begin after he served the minimum term on the rape and molestation convictions that had been imposed a year and a half earlier. Though Kondro was being spared the death penalty, county prosecutor Jim Stonier wanted to make sure he would never again be free or paroled.

At the sentencing, Rima's mother Danelle Kinne said, "I have been waiting fourteen years for answers, yet to finally have answers does not ease the pain. The fact that he deceived me for so long, all the while knowing the truth, leaves me with an inability to comprehend the monster that lies beneath the shell of a human being who sits here today."

12

INSIDE THE WALLS

The Washington State Penitentiary at Walla Walla, known as the Walls on the inside, is over a hundred years old and stands amid agricultural fields in a valley between the Palouse hills and the Blue Mountains in the southeastern part of the state, near the Oregon border. The original buildings were constructed of brick made from clay that was dug nearby; the thick stone walls were built of concrete and heavy stones brought from the Columbia River basin. The walls were designed to contain the worst of the worst, a function they continue to serve to this day, now enhanced with chain-link fencing topped with coiled razor wire.

It is the prison administration's aim to keep inmates as busy and occupied as possible—through exercise, through classes, and through work in the facility's various shops and maintenance departments. The original jute mill was transformed into a license plate factory in 1921 and currently produces more than two million plates each year. For those considered too dangerous for these activities or circulation within the penitentiary's general population, there is a maximum security unit where inmates are confined to their cells twenty-three hours a day, are given meals through a slot in the door, and are overseen by at least two guards whenever they are out.

This was Joseph Kondro's permanent home.

The inmate interviews that have been the most difficult are the ones where the media has followed me into the prison. While I was still in the bureau, CBS's *60 Minutes* arranged to have correspondent Lesley Stahl and a film crew follow me and my Investigative Support Unit colleague Judson Ray into the Pennsylvania State Correctional Institution in Rockview, near Penn State University. We were there to interview Gary Michael Heidnik, who was incarcerated for imprisoning and killing several women in the basement of his home in North Philadelphia. In this basement he had dug a pit, which he would fill with water and then place one or more women in and shock them with an electric wire. Thomas Harris used this aspect of Heidnik's crimes as one of the composite bases for his character Buffalo Bill in *The Silence of the Lambs*. The film version had recently come out and the media and public were clamoring for "the real story." Heidnik had tried to use the insanity defense, but I learned that at the same time as he was taking such pleasure in torturing women in his basement, he had made more than $600,000 in the stock market through his own investment strategies. He was executed by lethal injection in 1999, as of this writing the last person to face capital punishment in the state of Pennsylvania.

Despite our attempts to establish a rapport with him, Heidnik was cordial but wary. He had a kind of distant look in his eyes that I recognized from experience as advanced paranoia. He was already in protective isolation due to attacks from other prisoners, which only furthered the (in this case, realistic) sense that people were out to get him. This, despite the fact that he had a high IQ and had made a lot of money in the stock market without the benefit of a higher education.

He couldn't deny that he had held these women captive, but like southern slave owners trying to defend their indefensible institution, he insisted that he and the women were one happy family, celebrating birthdays and holidays together; that he gave them gifts and brought in food delicacies. He even mentioned the radio he bought to entertain them, which Jud suggested was actually employed to mask their screams.

Yes, he had had to beat some of them, he finally admitted, but it was for their own good, the way a parent might justify spanking a child. He had started his own church, and his ultimate plan was to use these women to populate the world with little Heidniks. It sounded bizarre, but he insisted he'd been serious about it. He went along like that, weird but calm, until I said I wanted to examine the problems in his background.

"Tell me about your mother," I said as I leaned in close to him.

That's when he suddenly lost it. He rose to his feet as if he was going to rip off his microphone and leave. I told him our research suggested that most serial predators like him had had either a severe conflict with their mothers or some tragedy where they had lost them. At that, Heidnik began sobbing uncontrollably.

I was able to get to him that way because of our thorough, preinterview research. I knew that, growing up in Cleveland in the late 1940s and early 1950s, Gary and his brother Terry were raised by a cold and emotionally abusive father who belittled him and threatened him physically, and an alcoholic mother who divorced the boys' father when Gary was two and Terry still a baby. They went to live with her, but within a few years her alcoholism forced them back with the hated father. She married three more times before committing suicide in 1970.

With cameras present, it takes a much longer time to develop a rapport with an inmate—generally, not because he is intimidated, but because he is so concerned with looking good on television and coming across as a victim. You have to move past all of that before you can get to the factual and emotional truth about the subject. There's also the simple problem of numbers. Back when I was doing interviews for the FBI, there would be only one or two people present in the room for the conversation. A television crew requires multiple people for setup and lighting and staging the room. If there are too many people or the room is too big, it is even harder to direct the conversation to where I wanted it to go. I figured it would take me at least an hour just to get Kondro to focus on me and my questions.

MSNBC executives expected me to get in the face of the men I would interview, to show overt disgust and contempt. This confrontational style might create tense and exciting television but would not make for a productive interview on my terms. This was almost a replay of our earliest prison interviews, when bureau higher-ups wanted to know why we were getting so cozy with killers and wardens questioned why each interview took so long. It can also be disconcerting for television viewers who don't understand why I'm being so "nice" and buddy-buddy with these evil people.

Perhaps the most-discussed exchange in the first season of the Netflix *Mindhunter* series occurs in episode 9, when Bob Ressler and my characters, called Bill Tench and Holden Ford on the show, interview student-nurse mass murderer Richard Speck at the Illinois state prison at Joliet. In an effort to get past Speck's contempt and get him engaged, Holden rhetorically asks him what gave him the right to "take eight ripe cunts out of the world."

It was actually pretty much like that in real life. We were in a conference room in the prison with Speck and a corrections department counselor and Speck was consciously ignoring us. I turned to the counselor and said, "You know what he did, your guy? He killed eight pussies. And some of those pussies looked pretty good. He took eight good pieces of ass away from the rest of us. You think that's fair?"

Speck listened to the conversation, then turned to me with a laugh and said, "You fucking guys are crazy. It must be a fine line, separates you from me." That was when the interview began in earnest.

The room in which the interview with Joseph Kondro was to take place was about twenty by forty feet. The crew numbered eight, employing three cameras. Then there were about half a dozen correctional officers to oversee the interview and make sure Kondro did not become violent, since I had requested that he be uncuffed.

When Kondro was brought in from his cell and saw all the people and equipment in the room, he looked somewhat shocked. He was large and stout, and looked strong. When we shook hands, his hand enveloped mine. It was impossible to have read his file and not think immediately of those hands beating and strangling young girls.

AS I WAS PREPARING FOR THE INTERVIEW AND GOING OVER HIS RECORD, I EXPECTED Kondro to react something like Charles Manson when he climbed up on the back of the chair to face Bob Ressler and me—a grandiose type who would want to dominate the encounter.

But instead of Charles Manson, I found in Joseph Kondro a compliant guy who seemed content to talk about his kills. My immediate task was somehow to get him to forget about the cameras and the crowd. This took some time, and he wanted me to know that he would not talk about any other case. He was a strong suspect in a number of other child murders and knew that law enforcement officials would have been delighted to pin another rap on him to get him into the death house.

But just because he was compliant didn't mean that this would be an easy conversation. Adhering to the strategy, I intended to carry out the discussion much as Bob and I conducted our offender encounters while in the bureau. You're looking for specific answers that could be beneficial to investigators. You don't pile on question after question, particularly if the subject is guarded in his behavior. At a glance, some of it may seem mundane, but every detail and behavioral indicator is important to someone like me, trying to open up and understand the mind of a particular violent offender.

I began easily, taking the high road. "I appreciate you taking the time to talk to me. What we're trying to do is really to educate the public, educate law enforcement, schoolteachers, to really kind of look into your background and see if there are any indicators early in your childhood and leading up to the crimes and discussing the crimes themselves, which we believe would be very beneficial. And what I would like to start with is to talk to you about your early years."

The first thing I brought up was that he was adopted. He confirmed that he had been—at around eighteen months—and launched into a detailed account of how his adoptive parents' families came over from Europe. Interestingly, I thought, he referred to Eleanor Kondro as his stepmother, which could be a sign of a psychological attachment disorder. He said that they didn't tell him of his own origins until he was about seven.

"How did you feel when they told you that?" I asked.

"I had a lot of mixed feelings about it," he replied. "You know, I have always wondered why anybody would give up their child."

"Like a sense of abandonment or something?"

"Yeah, a sense of abandonment. I thought that was one of the main issues to the way I was acting out, sort of acting out, as a child."

It's something like fishing. You throw out a line where you believe the fish might be and show them the bait.

In no way am I disparaging the blessing of adoption, but it is significant to note that a number of the most notorious serial killers have been adopted. These include Richard Ramirez, the Night Stalker; David Berkowitz, the Son of Sam; Kenneth Bianchi, the Hillside Strangler; Theodore Bundy, the Coed Killer; and Joel Rifkin, who murdered prostitutes in New York City and on Long Island. While the vast majority of adopted children thrive with their loving parents, I think if a young boy already has certain kinds of psychological issues or incipient antisocial personality disorder, the knowledge that he has been given up or "rejected" by his biological parents can fuel feelings of hostility, authority conflict, and negative behaviors. However, for these same reasons, this can be an easy excuse for a predator to "explain" his motivations and actions.

Kondro recalled that it was around the time he learned he was adopted that he started acting out. "I started to get very violent. Back then there was no mental health program in our school system and stuff like that. We just had a principal. I mean, all the principal would do was disciplining, you know, and that's where I started to, you know, act out in school and stuff, started to beat people up. I was a bigger child than everybody else."

Not surprisingly, many violent offenders were bullies or were bullied themselves as schoolchildren.

"You went to Catholic school?"

"I went to the Catholic school, yeah—private Catholic school—and I did really well there. I excelled at sports and I went to public school after I was in eighth grade, and that's when I started really doing drugs."

"What kinds of drugs were you doing?"

"Lot of marijuana, LSD, speed . . ."

I asked if, at that age, he tended to pick more on boys or girls.

"Just anyone around me."

"Were you ever molested as a kid?"

"Never was."

"At what point in your life were you developing [sadistic sexual] fantasies? Do you remember when that would have occurred?"

"Around the same time."

This was not unexpected. Of all the sexual predators I have interviewed over the years, a full 50 percent of them had their first rape fantasy between the ages of twelve and fourteen. Ed Kemper and BTK Killer Dennis Rader began to fantasize about crimes of violence before their teens. Kemper would cut the arms and legs off his sister's dolls, and Rader would draw graphic pictures of bound women being tortured. You have to wonder, what would have happened in any of these cases if someone had tried to intervene with them at that point.

Kondro described making girls strip and then "experimenting" with them, though he denied going as far as intercourse. He also talked about cruelty to animals—the key component of the homicidal triad—that began when the Kondros moved to Longview, Washington.

"I got into a group of kids; they were known as the neighborhood bullies. And one day this one kid that I was hanging out with, he said, 'Come on, I know where the cats are. Let's go kill them and stuff like that.' I wanted to. He took a baseball bat with him. That's when I witnessed my first killing of an animal."

Kondro would go on to use large pieces of wood to strike his animal victims in the head when they weren't looking.

"How did you feel?" I asked.

"I was kind of scared at first, and then I got excited. And then it was a chase because the police were out there. Somebody called in and we climbed up into a tree and we watched the police canvass the whole area."

His only concern was getting caught, but even evading the police was a thrill. There was no remorse for having participated in beating a defenseless animal to death. By this point we can see clearly how he grew up devoid of any feelings of empathy for the suffering of others.

"How did you get along with the opposite sex?" I asked. "I mean, you dated?"

"Yeah, I did," he said. "I did a lot of dating. I had a lot of girlfriends."

I still wondered what so many girls would have seen in him.

"Did you have a preferential age that you may have been dating, say, a teenager, but the fantasy was younger, a younger person?"

"Yes," he responded, and here was where it got really interesting to me. "As I got older, the girls would stay mostly the same age or younger than me, you know. Then the separation between ages started getting wider and wider, and pretty soon I found myself, you know, molesting kids."

"And so, what you are saying is that as you grew older, you were locked into a preferential victim. At what age did you prefer that group to be, were you looking at?"

"When I was going to school I just preferred girls who were younger than me, you know, a couple of grades younger than me."

"It sounds like it was an obsession, you know—obsessive thoughts. Were you able to control these obsessive thoughts that you had?"

"I thought about it all the time," Kondro conceded.

I asked if he drew pictures like Dennis Rader.

"No, it was just in my head. I mean, I just thought about it all the time, thought about taking people out and raping and killing them—

in seventh grade." These types often play out the ideal script of fantasies in their minds. The actual crime seldom completely lives up to the script.

"So there was a hell of a lot of anger. Where do you think this anger . . . just going back to seven years of age when you found out you were adopted, that was almost like abandonment to you, do you think that . . ."

"That was part of it, but my adopted parents, they were very, very controlling people. I remember telling my mom I didn't want to go to church anymore, and she made me do that. They would force me to do stuff. And they would yell amongst each other a lot, abusing each other mentally. They would always yell at each other."

This, he felt, led him to do "a lot of deviant stuff with the kids around the neighborhood."

"What was that?" I asked. "What do you mean by deviant stuff?"

"Well, you know, I either beat boys up or I take the girls off and make them strip down for me, you know, play our little games. But pretty soon I noticed that all the parents in the neighborhood wouldn't let me play with their kids anymore. So in an effort to move closer toward the job, my stepdad moved us to Longview, Washington. Now, I don't know if that was based on what I was doing or maybe he just wanted to get closer to his work. It could have been a combination of both. But I was never told about it."

One of the things that is interesting and pretty consistent among serial predators is that two emotional concepts are constantly warring within them. One is a feeling of grandiosity and entitlement. The other is a deep-seated and pervasive sense of inferiority and inadequacy. I certainly saw this in Joseph Kondro, and it permeated nearly every aspect of his character and outlook.

His attitude toward his parents was representative of this dichotomy. After essentially telling me how dysfunctional his family life was and how his parents were always yelling at him and emotionally abusing each other, he added, "But they were really good people. I was their only child. They got me everything that I wanted when I was a kid. My dad taught me the value of money. He put me to work at the age of twelve and I had my own little landscaping business. Yeah, they were good people."

As far as "the value of money," this was a man who was frequently out of a job and had figured out ways to sponge off his friends, ex-wives, and girlfriends for extended periods. With guys like this, the concept and the actuality are two different things and they don't see how it relates to them.

"Finally," I continued, "at what point really kind of tipped you over, when you were beginning to act out the fantasies?"

"Maybe twelve or thirteen years old. One night there was this girl that was working in a local neighborhood store, and I fantasized about taking her out and raping her and stuff. And I put together this rape kit and I went to the store after they shut down at eleven and she was just locking up. And I asked her for a ride and she said, 'Sure,' and she got me into her car and I pulled the knife on her and stuff and took her out to the Mount Solo area. She cried and stuff like that and 'Please don't do this, please don't do this!' and I—I didn't do it. I didn't go through with it."

"So you stopped? And you felt sorry for her?"

"Yeah, I did."

"So it did touch you at that time? You had some feeling?"

"Well, it was my first act, you know. I didn't know what I was really doing."

"So is it fair to say that in the fantasy everything kind of was perfect, but in the reality, when things didn't go as planned—you thought you had a very good plan, but you didn't expect her to cry, you didn't expect her to react that way?"

"Yeah. Later I overcame those, you know, feelings." In other words, he learned to detach himself emotionally from his victims, so their pleas or suffering didn't affect him or stop him from doing what he intended.

"Did she report to the police or anything?"

"Yeah. I went to court over it, but the lawyer got me off."

And as a result, at least two innocent young girls died.

13

"THE CONVENIENCE OF THE SITUATION"

We had gotten past Kondro's childhood and formative years, and we were heading into the one question about him that interested me above all others: Why would he risk abducting and killing the children of people he knew well? For anyone smart enough to pull off multiple crimes, this struck me as extremely high-risk behavior.

And bound up with this question was the more basic psychological one of why he chose preadolescents as victims. My instinct was that his feelings of personal inadequacy drew him to victims with whom he could feel equal. We're all familiar with milder versions of this, such as the guy who graduates (or doesn't) from high school but keeps coming back to hang around with younger kids who will look up to him when his peers won't.

I asked, "Why were children your target? Why not, say, a woman, eighteen years old? I mean, why a child? You've had plenty of time to think about that. What's going on in your head that you fixated there? What were you trying to get out of this?"

Kondro's answer was stunning in its directness and simplicity.

"I think it's just the convenience of the situation. You know, children are very trusting and stuff, and I was very close to their family members and, you know, I just, I played on their trust."

"So it was really, basically, it was an easy target for you."

"Yeah, it was an easy target at that time."

His background showed that for a variety of psychological reasons, he preferred young girls, and I was still convinced that had primarily to do with his personal sense of inadequacy. But the *strategic* reason for targeting young girls was simply that they were easy prey, like the lion that tries to pick out the most vulnerable antelope at the watering hole. An eight- or twelve-year-old is not going to be able to put up the same struggle as an eighteen-year-old.

But before he started killing, wasn't he worried that any of the girls he molested would go back and tell their parents?

"Yeah," he conceded. "I molested one girl and she told her mom and her mom confronted me and . . . She didn't do anything about it. We just discussed the matter and then she left. I was expecting her to go to the police department, but she didn't do that. She decided not to report it."

Consider this statement in relationship to the incident in which Kondro threatened the teenage girl with a knife on Mount Solo, for which his lawyer was able to get him off. Many parents are loath to go to the police for fear that everyone in the community will know their child has been molested, and they do not want the stigma or for the child to have to testify in open court. Kondro was starting to understand the "ground rules" and exploit them to his own advantage.

"You think there could have been anything in your life at that point—now we are talking about early childhood—could anything have been done to prevent you from stepping over the line, committing any violent crime?" I asked.

"No, I believe that most of the molesters, I believe that it's put in our genes, you know, down the family line. I have another family member in another prison across the United States. He is also a molester. So I truly believe, just like alcoholism and drug addiction and stuff, this is an epidemic that the government agencies have to really take a good look at, because they do a lot of research, because I think it's in our genes."

At first this sounds very analytical and high-minded, as if Kondro was taking the big-picture view of the problem of which he was one example. But when you get down to it, it is just another way of not taking responsibility for his crimes. It's like alcoholism; it's like drug addiction. It's hereditary: *I was born to molest and kill and there's nothing I can do about it, so it's up to government agencies to do the research and figure it out.*

Nonsense. An alcoholic is not absolved of beating his wife or hitting a pedestrian with his car simply because he is drunk. Kondro may have had a strong urge to rape and kill for his own satisfaction, but he was not compelled to, nor is anyone else. He was too concerned with the "safest" way to commit and cover up his crimes for them to be "irresistible." This is all about the choices each individual makes.

When she was interviewed, Kondro's daughter Courtney also rejected the premise. If such behavior had been genetic, Courtney reasoned, she would have the same urge to kill as her father, yet she had never had the slightest urge to hurt anyone. She also calls her father's actions a choice, not an excuse.

I followed up by asking Kondro, "Would you say you have an addictive personality?"

"Yeah, very addictive, more 'now or never,' you know."

"More now? Even in here? In what way?"

As soon as he saw what he could get away with and what I would challenge him on, he often modified his statements.

"I don't know. It's not really addictive, just maybe a compulsive personality, you know. There is nothing to do in prison. You have to do your own time and you have to make that time count. And most people do it by compulsively cleaning our cells and doing programming or play sports or, you know . . ."

"What do you do?"

"Well, I clean my cell a lot. I am just like all the other inmates. I like to keep my cell really clean."

First of all, he's not just like all the other inmates, though he'd like to think so. He's a child molester and killer, which puts him at the very bottom of the prison pecking order. Second, being incarcerated, he no longer has control over his life. He is frustrated that he can no longer act out his fantasies. One way to have some type of control is to perform obsessive-compulsive acts like cleaning and recleaning his cell—the only domain over which he and the other inmates have any control.

I also thought this could be an opening to what he had been thinking at the time of the crimes. "Do you think that was the time in your life when you felt like you were out of control, you know, with what was going on—in relationships, your employment—that this [the crime] would be the way, the one time where you could commit an act and be in control, be the boss?"

He took the bait, laying his behavior off on the influence of others, "That's part of it. I found out that a lot of people in my life were trying to control me. You know, all of my girlfriends would always want me to change, my mom wanted me to change, my dad wanted me to change, all of my friends thought I should change. You know, I was an alcoholic and a drug addict and I just, I was abusing everybody around me."

Again, he seemed to be showing some insight and taking some responsibility by admitting he was abusing everybody around him. But really, it was everyone else's doing. Rather than wondering if the fact that everyone he knew wanted him to change meant there was some wisdom in that, he saw it as exterior control. His way of escaping this oppression was to abuse drugs and alcohol, which would also have the effect of contributing to the lowering of his inhibitions prior to perpetrating his crimes. This also indicates his narcissism and lack of empathy for anyone other than himself. Not only that, but he was an angry drunk. He has a lot of pent-up anger at the world, so he will show no empathy or compassion for his victims or responsibility for his crimes because he feels that he, in fact, is the real victim.

Dennis Rader told me that his BTK reign began because he lost his job. "It was all because of my getting laid off at Cessna. I wasn't having any sexual problems with my wife or financial problems. It was all because of unemployment. It didn't seem fair. I really loved that job."

I found this a pretty interesting statement to analyze. On one level, there is no way being laid off from a job ever causes a reasonable person to terrorize, torture, and murder four family members and then want to brag about it afterward. That is just an excuse employed to mitigate his sense of responsibility. On another level, though, this speaks directly to Rader's psychological makeup, and I don't just mean his longtime fascination with ropes and torturing women that finally evolved from thought to deed. What it shows is Rader's profound narcissism, that something unfair was done to *him*, never mind that he responded with an action that was unfair to his victims and their survivors, countless orders of magnitude greater.

Dennis Rader had a wife, two children, and a government job. At one time he was a Cub Scout leader and president of his church. But the bondage and torture fantasies that he conceived, planned out, sketched on paper, and then went out and executed were by far the most important thing in his life.

Though their murderous fantasies were equally depraved, we see the difference between Kondro and Rader's signature and M.O. Kondro wanted to rape and kill with the least amount of effort or resistance from his victim. For Rader, the sublime satisfaction came from sadistic pleasure he derived from watching his victims as they knew he was going to kill them. Once he had his victims bound and gagged, he had no interest in inflicting physical pain as most sadists do. Rather, he gloried in the power of life and death and the victim's anticipation of death. For Rader, the victim was a lead actor in his hideously conceived dramas. For Kondro, the victim was merely a prop, to be discarded when he was finished with it.

I knew from Rader's background that, like Kondro, he had started out on animals and I confronted him with that fact. "Tell me about the animals, Dennis," I said.

His expression grew grim. "I know where you're going with this," he said. Of course, he did. He had read our books and knew it was part of that homicidal triangle—along with bedwetting and fire-starting. "But I never killed any animals. I would never have done that." I found it interesting that he could practically brag about what he had done to human beings but was too ashamed of mistreating small animals to admit it.

What this statement actually demonstrates is the one universal among all serial killers and violent predators: other people don't matter, they aren't real, and they don't have any rights. It is sociopathology carried to its farthest extreme, and for killers like Joseph Kondro and Dennis Rader, it defined their interactions with the world.

IT WAS TIME TO MOVE KONDRO ON TO THE RIMA TRAXLER CASE.

"Well, I met Rusty, Rima's stepdad, in school," Kondro began. "And then I met Danelle one night at a bar and Rusty introduced me to her and we just became friends. You know, Rusty and I were friends from way back. I was working at this local smelter and I ran into Rusty again—he came back from some job in Nevada and we started hanging out again."

Now, the first thing that should occur to you about this brief narrative is how absolutely ordinary, matter-of-fact, and almost banal it sounds. Then you remember that Kondro killed the stepdaughter of his "friend from way back," and you realize once again the searing reality: Predators may look and sound and often act like we do, *but they don't think like we do*. Their logical process is completely different.

If Kondro's previous statement described a life that was prosaic, his next demonstrates what actually constituted everyday life for him and his social circle:

"Well, at the time, we were doing a lot of drugs, a lot of drinking, a lot of parties. Cocaine was the main drug that we were dealing. Rusty had lost his job or Danelle had kicked him out or something like that and he was living on unemployment benefits, renting out this house, and he couldn't keep up the rent, so he asked me if, you know, I would mind him moving in with me, and I said, 'Sure, you can move in with me.' And I was seeing my ex-wife Julie at the time, so I was never there. And then one day we just got into this conversation and he told me the password that they had established with Rima."

"What was the password?"

"It was *unicorn*, and one day I just seen her walking down the street and I went to the store and when I came back, she was still walking down the street. I pulled over—it was just an impulse—I pulled over and picked her up." This wasn't an impulse; an impulse is fleeting and sudden. This was a force that had always been there with him. It was a crime of opportunity.

"Then you said the password?"

"Yeah, yeah. I said the password and she got in. She got into the truck with me and then I drove out to the Germany Creek area." Notice how matter-of-fact he is when he gets down to talking about his M.O.

At a certain point in conversations with killers, it always comes down to this:

"What happened, Joe, on that day? What finally was the stress, or what gave you that push that, yeah, you are going to commit a crime and it's going to be this child, you know, Rima?"

"Well, I zeroed in on Rima because, first of all, she trusted me, and at that time in my life, you know, I was attracted to young girls. So I decided to go with her, and Rusty gave me the password, and like I said, I was going to the store, and the way back from the store I had seen her walking down the street still and I said the password and she jumped in my truck. So I just took her down to my house and I told her to stay in the pickup, and I went in the house and called work and said I wouldn't be in that day. And I drove her out to Germany Creek and raped and murdered her."

"Did you assault her in the car or out of the car?"

"I took her out to the old swimming pool that I knew of and she was just standing there looking at the river and it was—it was a swift river—but there was a pool there and she was standing there looking at that, and I just hit her in the side of the head with, you know, my right fist, and knocked her unconscious. Then I raped her. She started coming out of her unconsciousness when I was raping her, and I just started to strangle her."

"Manual strangulation—your hands?"

"Yeah."

"Was she facing you or was it from behind?"

"She was facing."

You can tell by how an offender manually strangles his victim if he genuinely intends to kill, and if there are any reservations, moral hesitation, or empathy. Here there was none. Kondro and Rima were face-to-face. This was a man she trusted, yet he had no compunction about strangling her as he stared into her eyes. I put myself in the victim's place, and I think that the last image I will ever have is that this man whom I trusted with my life is taking it and doesn't even feel the least regret.

"Facing her—was that hard?" I asked. "Was that hard in retrospect."

Not in the slightest, apparently. "At that time, I was committed, you know. I had already made up a decision that I was going to kill her, way before she even knew it, when she got into my truck."

"There was no talking between you or anything like that?" I clarified. "She really didn't know what hit her? I mean, you rendered her unconscious and then you had sex with her after death?"

He corrected me on the detail of when he had raped Rima. It was as straightforward and procedural a recounting as if he were describing the effort to change a flat tire. Yes, he had hit her hard enough to neutralize her, then had had sex with her when she was unconscious. He strangled her, and as she regained consciousness and gasped for air, he continued sexually assaulting her. Then he dragged her over to the nearby creek and forced her head underwater. But when he raised up her head again, she was still alive and gasping for air. So he grabbed a rock about the size of his hand and bashed it against her head until she died.

As I listened, I was conscious of not changing my expression, but my blood was boiling and I was thinking to myself, *If anyone deserves the death penalty, it's this guy.*

This narrative takes my breath away—not only for the sheer depravity and evil of what Joseph Kondro did to this young girl, but for his apparent attitude toward the act and to her. Though he didn't get off on watching his victims' emotional anguish in anticipation of death, he was so clearly dispassionate and uncaring about the suffering of those who knew and trusted him that in his own way, he is just as cruel and narcissistic as Rader.

This confession was too raw for the television broadcast, but I had gotten him into the zone, the same way McGowan had relived his crime. He had never previously described what he had done in this detail. It didn't seem to matter that the guards and production crew were listening, Kondro had showed what he was really all about, how he proceeded to get what he wanted, regardless of the consequences to others.

Most predatory killers, particularly the sadistic ones (those whose primary emotional satisfaction comes from inflicting physical and/or emotional pain and making others suffer helplessly), need to depersonalize their victims, to be able to treat them like objects. This is harder when they know the victim, though clearly, this was not the case with Kondro. He knew Rima well; he was a longtime friend of her stepdad as well as her mom. He had seen her grow up and knew that she trusted him. He liked her and had no gripes against her. There is no way he could depersonalize her in the traditional sense. And yet he was able to decide to rape and murder her and carry it out with seeming methodical dispassion. This was what made Kondro unique to me, and why I wanted to understand him.

There are some predatory personality types who also kill people they know and who are close to them, but there is generally a different type of psychodynamic involved than was the case with Joseph Kondro and the crime has a different behavioral presentation. An offender who kills someone close to him is usually motivated by a great sense of perceived betrayal, revenge, or anger, often fueled by jealousy and outrage. We saw this with the O. J. Simpson case, in the murder of his ex-wife Nicole Brown Simpson and her friend Ronald Goldman. In a situation like this, we would expect to see behavioral evidence of "overkill"—much greater harm and more violence to the victim than is necessary to cause death. A typical behavioral pattern of overkill is multiple stab wounds in a tight pattern on the neck or chest, and severe damage to the face. It is a murder in which the primary goal is punishment fueled by rage. Ron Goldman just showed up at the wrong time and place. Simpson had no agenda with him other than to neutralize an unexpected threat. Nicole was the one singled out for violent punishment.

That was not the case with the Rima Traxler murder. Kondro had no rage against her. He had no reason to want to punish her or her parents. He never even expressed any dislike for her. It was just something he decided to do, practically on the spur of the moment, because it was easily doable and would give him pleasure and satisfaction. He decided to rape her, and it would be easiest to rape her if she was not resisting. He didn't get off on her struggling or suffering. Then once he had done what he set out to do, he had to kill her so he could get away with it. If she died by strangulation, fine. If not, the next option was to hit her with a rock. It was detached and coolheaded, like killing cows in a slaughterhouse.

"How did you dispose of the body?" This is another of the questions that are always highly significant to me.

"Well, I was pressed for time. I didn't go out and dig a hole or anything like some of these other murderers do, but I just took her over to this big old broken log that was sitting up against this cliff and threw her behind the log and pulled out a bunch of ferns that were around the area and threw them on top of her and disguised the area as best I could and left, took all her clothes with me and drove over to the Longview–Rainier bridge and tossed some out into the river."

This sloppy attempt at concealment told me that Kondro was probably under the influence of drugs or alcohol. We knew from the police interviews with some of his friends and associates that he had said the key to not getting caught for murder was to make sure the police couldn't find a body. This rushed, careless attempt at body disposal easily could have gotten him caught. He was just plain lucky that it didn't.

He admitted as much to me. "I was really surprised at that because when a body decomposes, it stinks, you know. And that was a popular swimming hole. I am surprised nobody found her."

This is the part that most of us will find utterly unfathomable. In her celebrated and controversial book *Eichmann in Jerusalem*, philosopher Hannah Arendt wrote about the Nazis' "banality of evil." Here was a perfect exemplar of the phrase, only a few feet away from me. I asked Kondro, "So, when you were done killing and then you went home, did her mother reach out to you—for help?"

"Yeah. I went back to my girlfriend's house at the time and she was cooking dinner and I had a couple of chores that I needed to do around her house, and about—I don't know—about six at night, it was dark out—I remember there was a bang at the door and it was Danelle, and she was asking if Rima had come over and we said no, and she started crying and she used my phone to call the police department. So my girlfriend at that time, she told me to go with her to make sure, you know, she was all right. And we went directly over to another person's house and she asked them and then we got back into the car and we went over to Rusty's house and there was cops everywhere at Rusty's; they were tearing the place apart. They thought he was responsible for the disappearance."

His close friend becomes the primary suspect, which is fine with Kondro because it takes the focus off him. Rather, the police probably overlooked Joe to begin with because he helped the victim's mother.

"Did they ever go after you, interview you?"

"I don't believe that they ever talked to me back then. They, Lisa Snell, I think her name was—a reporter—went to talk to Rusty and the police had a few interviews with Rusty and he failed the lie detector test even . . ."

Examples like this are one of the reasons I've never set much store in polygraph exams. It is largely ineffective on suspects who already have a criminal history and who may currently be involved in other crimes. They believe in their own warped minds that their crime was justified, or they were entitled to do it. Or, as several serial predators have told me over the years, if you can lie well to the police, how hard is it to lie to a box?

". . . They never ended up finding her [body] and stuff, and they just kind of faded away."

"Did you ever think you would get caught? You felt pretty good that you wouldn't be caught?"

"I felt pretty good that I wouldn't."

"Why is that?"

"Because their main focus was on Rusty, you know, and I just let that ride."

"Did you ever go back to make sure the body remained concealed?"

"No."

"Never did?"

"No, you never go back."

One of the ways we analyze UNSUBs is whether there is any evidence they have returned to one or more crime scenes or body dump sites. It isn't even necessarily that we are going to surveil a crime scene in hopes of catching the offender, but whether he returns or not—either way—gives us a lot of behavior to work with.

There are two main reasons killers return to these scenes. One would be classified as signature behavior and the other as modus operandi—the psychic versus the practical aspect of the crime. The first is to relive the thrill and emotion of the crime. We have seen a number of offenders come back to masturbate on or around the victim's body. We have even seen actual necrophiliacs who returned to have sex with their recently deceased victims. Ted Bundy was one of those. Clearly, Kondro did not fit this profile; once he was done with a victim, he was done, and blithely went about his business.

The other reason is defensive: to make sure the body remains properly hidden and will not be discovered by the authorities or casual passersby. Kondro made it clear that he did not participate in this type of behavior, either. He thought that going back to the crime scene or dump site made it more likely he would be found out. Why was this? What separated him from the thinking of any other offender who didn't want to be caught?

The fact that he was already associated with his victim. Unlike other predators who target strangers, Kondro knew he would come within the circle of potential suspects and that his movements about the area might be observed or scrutinized. He was uncomfortable with the fact that he was "pressed for time" after Rima's murder, but the police seemed to be focusing on Rusty, so better to leave well enough alone.

That Rima's body was never found, even after Kondro agreed to cooperate in locating it, is unusual, but it certainly worked to his advantage, as opposed to the ongoing agony it created for her family.

I said to him, "Her mother to this day holds out hope that her daughter is alive, even though you said you were responsible. Because she has never found her, they hope that she would still be out there. But there is . . . there is no hope."

"No, there is no hope. She was . . . she was dead when I concealed her body."

"If you could say something to the mother, what would you tell her?"

"That's kind of hard," he said without much pause for introspection. "I don't know what I would say to her. What do you say to somebody that, you know, goes out and kills children? There is really nothing to say. I mean, the act speaks for itself."

14

"THERE WERE VICTIMS IN BETWEEN"

Joseph Kondro had been convicted of two murders, one in 1985 and one in 1996. From all that I knew about sexual predators, it just didn't ring true with me that he could lie in wait for eleven years.

I brought up his image of an alligator waiting for long periods of time below the surface until he got "hungry." I asked him, "Can you really sustain yourself that long? Usually it is a fantasy that sustains it. I mean, reliving the crime over and over in your head?"

Again, matter-of-fact: "That was part of that. But let's get one thing straight. There were victims in between these. I mean, there were molestations that were never reported, you know."

"I am really glad you said that," I responded, "because when I looked at the case, I said there is no way that he can go ten years based upon that type of case, that you committed initially and then the one later on. The people I have talked with—some of them will take maybe mementos, things that belong to the victim or [save] newspaper articles, but the urge, when you have this obsession you need an outlet. So there were these other cases, and many went unreported?"

"Yeah, there were a lot of unreported cases."

"Were the victims always friends, friends of yours?"

"No, they were just like—like I said, I liked to party a lot back then, and you know, there was young girls always there, you know, fifteen, fourteen, sometimes thirteen. And stuff would be at the parties and they were just party girls."

"Why is it that you didn't kill those victims?"

"Probably because I wasn't done with them yet."

"What did you want to do with them?"

"Just keep molesting them."

"Was it part of the fantasy to keep them alive for a period of time? What would be the ideal situation for you to have a victim? To play out the whole fantasy, from the beginning to the end, regarding location, time, whatever?"

"My fantasy was just to murder—I mean rape and murder my victims. That was the whole fantasy part. My murders, you know, they were like the high point of the whole thing. In the past, I molested these people, I mean these children and stuff like that; maybe I didn't kill them because I liked them, you know."

But then he went on to say, "For me, I don't look at it as a fantasy. I was looking at it, during the time I was free, I was being, doing the act—you know what I mean—as a part of my life. It was a part of my lifestyle."

Though his preference for targeting girls he knew well for the ease and convenience of the crime was particular to Kondro, in other ways he conformed to a violent predator personality type. By that I mean his hunger for the sensation would be ongoing. Therefore, if it turned out that the four assaults and two murders for which he was behind bars were his only violent crimes, he would be an extreme exception that I would want to understand and place in the context of predator typologies. On the other hand, if there had been unknown and unsolved assaults between the ones he had been charged with, that would confirm what we knew about this kind of sexual predator. Either way, getting an answer was extremely important.

"I have read police reports where they thought you were good for maybe seventy victim molestations and whatever, even possibly other homicides," I said. "What do you know about that?"

"Just what I have read."

"So the numbers are an exaggeration, when you say like seventy, that you are good for seventy other cases?"

"I can't answer that."

"You don't want to answer it, or you just can't?"

"Yeah, I don't want to answer that." Because that might mean still having to face a death penalty. That answered my question. Kondro was not the sort of guy—and there are plenty of them—who, once caught, want to enhance their reputations by claiming more attacks or kills than they actually had. For Kondro, killing was too mundane to care about that sort of status. So if he wouldn't respond directly, it was because there was more to tell.

I asked him if he had ever gotten involved with a woman so that he would have access to her young daughter. This, unfortunately, is not uncommon among sexual predators of a certain type. He denied it, but I remain highly skeptical. What was pervasive throughout the interview, though, was the sense I got that he really didn't think he was responsible for any of this. He kept committing crimes because the police were incompetent in not arresting him. He escaped the death penalty because the victims' families wanted to make a deal with him. In his mind, other people always initiated everything. Even having the Traxlers' secret code word with Rima—they just gave it to him; he didn't ask for it. But once he had it, he felt free to use and abuse it.

"In my case, they let me make a deal, you know. They had me on the table for the death penalty and the prosecutor was pushing for the death penalty. So they went to the families, and when the case was finally over, you know, I made that deal with them. And that was a family member's decision to do that; the prosecutor went to them and they talked about it. And I feel that there is no justice in that, you know. They should never have given me a deal."

See what I mean?

"They should have given me the death penalty," Kondro continued. "I believe that I should have died for these crimes, you know. And—I don't know—I just feel . . . my victims didn't get any justice. They are dead; I am alive. They are rolling over in their graves. One of them hasn't even been found yet, and, you know, I just believe that the families—even though they were trying to do right by trying to find Rima Traxler's burial spot, they were thinking of the best interests of Danelle—in my heart I don't think that was right. There was no justice to it. They let their kids down."

In my heart I don't think that was right. Can you believe this guy? I had trained myself over the years not to react with my true feelings, to keep them in check, but sometimes it is difficult to keep your composure when you hear such outrageous statements. After all, he himself had made the deal that saved him from death. It would have been so easy, but so counterproductive, to give the MSNBC executives the kind of reaction they wanted.

He went on with his facile moralizing: "You know, if it was my kid that was killed, I would want that guy on death row. I would never ask the prosecutor to make a deal with him. That would be the last thing in my mind. Kill him, you know, get him out of this planet."

I almost asked Kondro why, then, he didn't just plead guilty to capital murder and conform to his own sense of justice. But I decided it was better to just let him keep talking.

After he lamented how unfair this plea bargain he was offered was to his victims, I asked him, "So [accepting] the plea bargain wasn't because you were trying to escape the death penalty? It wasn't for that reason?"

"Nobody mentioned plea bargain to me. I came up with that idea myself." And how appalling he felt it was that they accepted it!

"Were you afraid of death?"

"No, I am not afraid of dying today, you know." *Of course not, once execution had been taken off the table,* I said to myself.

"You were concerned about . . . you didn't want your own children to testify?" I asked this question knowing the plea bargain had nothing to do with his children—only perhaps his image of himself as a father. He hadn't exactly ever been Father of the Year.

"Yeah, it really bothered me because they were witnesses for the state, and the more and more I thought about the stuff, I didn't want my children [to have to testify]. They were young at the time—fifteen, fourteen, something like that, sixteen—and I didn't want them getting up as witnesses for the state and testifying what happened that morning and then going through the rest of their lives thinking that *Oh man, my testimony might have put my dad on death row,* and stuff like that. You know, I just didn't want my kids involved in any of that at all."

Even though he didn't even know their exact ages. Plus, he just got bored cooling his heels in the county jail while the prosecutors gathered evidence and considered what to do about him. "I just got tired of being in the county jail. I was in the county jail for I don't know how many months—twenty-eight months or something like that—in solitary confinement, you know, because they can't put me in general population. The whole, the jailer and the community, you know, wanted to kill me. So, I just wanted the whole ordeal over with."

Despite the deal to give them details about where he left Rima's body, he couldn't find it. "Yeah, that was like, thirteen, fourteen years that went by and that river, you know it rose and settled fourteen or fifteen different times. Myself, I was brought out there—I wasn't even sure that was the right place until I started recognizing landmarks. I drew them a picturesque sketch of the area that the place was, and the police officers, the detective, went to neighborhood people out there at Germany Creek and they went to show where the swimming hole was. So when I got down there, I started recognizing landmarks and stuff and I was sure that was the place."

He seemed more apologetic about his inability to locate the body than about the rape and murder itself.

OVER MY MANY YEARS OF OBSERVING AND INTERACTING WITH SERIAL KILLERS, I'VE found that a large percentage of them are abnormally fixated on their mothers—usually negatively, like Kemper; sometimes positively; or a confused mix of both, like McGowan. I wanted to see if Kondro had any more feeling for his mother than he did for the others in his life, and whether her influence had actually impacted him one way or another. I asked him about his mother's death, which occurred around the same time of year as his murder of Kara Rudd.

"Yeah, my mom, that was a big loss to me and to my stepdad," he conceded. "After my mom died, he completely lost it—in a way I had to put him in a home."

"He went downhill?"

"Yeah, then his brother came and stepped in and we decided it would be best to get rid of the house and stuff because I didn't want to live there anymore, you know. I was staying with them because my dad had diabetes and Mom was ill and she asked me for help, and so . . ."

I wanted to see if I could get him emotional; he certainly hadn't been very emotional about his crimes or his betrayals of friendship and trust. "You really loved your mother?"

"Oh, I loved her, yeah."

"So that was a crushing blow?"

"Yeah, it was a pretty traumatic part of my life. I was the one that found her, and there was a friend that I called, and she came over and . . . you know, I didn't know what to do. I was, 'You know what to do. You call the police department and have them come in?' She says, 'No, I will take care of the whole situation.' So she took care of it for us and they called the coroner and the police department knew that she was terminal, and that next morning I went in [to see her] and she was dead."

"From what I read, she was a wonderful woman."

"Yeah, she was a beautiful woman. She had lots of friends."

"Did that take you down, on the downside emotionally?"

"Yeah, it did. It took me down emotionally." I thought maybe I was finally getting to him, but his words and the tone of his voice were in such opposition to each other that I could see he was just going through the *motions* of feeling rather than the true *emotions*. And then he instantly switched gears: "But you know, I am a survivor. I needed to get on with my life. And I didn't have a place to stay in. Janet and Butch offered me a place to stay. I got a settlement [from his mother's estate] and I had a lot of money at the time and they offered to let me live with them, so . . ."

This man was revealing himself to have no meaningful emotional connection with anyone. It helped explain why murder was so easy for him.

"You went through that money fast?" I said.

"Yeah, I bought a couple of cars and some painting equipment. I was in . . . at that time, I was, you know, painting residential houses and stuff."

"Then?"

"Then a lot of drugs and alcohol."

"Drugs and alcohol. We will go back to that alligator now. It sounds like it was 'feeding time.' What do you think? Was it a group of precipitating things? I mean, a lot of it—the depression and you are angry and here it is, this is your friends together, now there is their daughter. Because you're involved in drink and drugs, is that going to be the cause for you now to perpetrate this crime, would you say?"

"I am not going to blame here everything that I did on drugs and alcohol. That was just something that I did as a pastime, you know. But I think really deep down, when you get down to it, the Kara Rudd murder was more of a revenge thing than it was like acting out my fantasies. That was part of it, too, but you know, I was just looking for revenge."

"Tell me about the revenge."

"There was a lot of phone calls being made about me, and Janet kicked me out of the house unexpectedly one day and I started blaming the girl [Kara] for it and I thought, well, you know, I have done so much for you guys, I bought you a car and all this other stuff and done all this other stuff and you kick me out of your house like that. But I can see, in her [Janet's] defense, I can see, you know, I was out of hand. I couldn't see it myself because I was the alcoholic, the drug addict, and stuff. I was just having a good time, but she had had enough of it. She had enough of that behavior, so she decided that she was going to kick me out and she did. So then it became a revenge thing. One day I was going, and I see the kids walking down the street and the morning routine was to ride the city bus to school, and I offered them a ride to school."

"Were you ready then to get them? Or was it just a practice run?"

"Well, it became a practice run later on. I took them to the school and dropped them off and stuff and then Nicky went into school, went off and played with his friends and stuff, and the girls asked me—you know they wanted to skip school that day. So I took them out to the river, to this abandoned house and . . . Yeah, that was a practice run. Then I brought them back about an hour later to school; they missed first and second period. But nobody said anything, the school didn't call the parents, they just missed class."

Again, it was everyone else's fault that he was not caught or held accountable.

"How many days later did you . . ."

"Oh, I can't remember, maybe a month, a month and a half."

A month and a half later he was out of money and thrown out of his friends' house. "It was about a month and a half," I confirmed. "And she was close to you, right? I mean, she called you Uncle Joe."

"Yeah, I knew Kara for a while. We even played cards together and stuff. I would let her win."

"How would you describe her as a kid?"

"She was very energetic. She liked bringing animals home. But she was very defiant, too, when she believed in things, you know, when she believed in something she was very defiant. She stood up for what she believed in. You know, you got to admire that in anybody, you know." Yes. Kara represented everything Kondro wasn't.

"So, this crime was a revenge and it hurt the mother?"

"Yeah, that's what it came to." Again, another rationalization. This wasn't about revenge. He would have committed the crime whether he had any beef with Janet or not. He had already shown us that.

We all feel the urge to exact revenge at one time or another, but most of us are able to suppress and control those impulses. Actual revenge murders tend to be one-offs rather than serial crimes. These types of murders have specific indicators and fall into one of two categories: retribution against individuals the killer feels have hurt or offended him, or retaliation against entire communities, such as school shooters who feel they have been bullied or disrespected. As a rule, though, while sexual predators tend to be acutely sensitive to any perceived slight or insult—while giving no mind to the feelings of others—they don't generally have revenge as a motive; they don't need it. They are already bound up with their own deadly obsessions, as we clearly see in Kondro's case.

Serial killers who claim revenge as a motive are usually manifesting some form of emotional displacement. Richard Laurence Marquette, whom I interviewed in the Oregon State Penitentiary in Salem, where he was serving a life sentence, had had an unsuccessful experience trying to pick up a woman in a Portland bar. From that he perceived that all women were rejecting him and proceeded to take his revenge. He picked up another woman, raped her, strangled her, and dismembered her body in his shower. After his conviction and eleven years as a model prisoner, he was released on parole in 1973. This case was on my mind as I evaluated Joe McGowan's case.

Two years after his parole, Marquette picked up another woman in a nightclub. He invited her to his trailer home, which was about a hundred yards from the bar. There he sliced the head of his penis, unknown to the victim, before he had forcible sex with her.

"Why?" I asked.

He said he wanted to believe and feel that the victim was causing his pain, which I interpreted as his way of justifying his crime. After he had sex with her, he strangled her and pulled off her fingernails with pliers. Internally I had to keep telling myself to relax, breathe normally but deeply, and not show any negativity either on my face or via body language. But believe me, it was a struggle.

Though I didn't believe Kondro had undergone a similar thought process, I continued this idea with Kondro to see where he'd go with it. "This would be what—taking something, like a possession, a close possession away from the mother?"

"Yeah. That morning when I woke up I knew what I was going to do. I was completely aware what I wanted to do, and stuff, so I drove to the school . . ." And again, he related the encounter with the two girls.

But then right after telling me that he woke up knowing what he was going to do, he went back to his original narrative and said, "Yeah, this all happened just by chance. In the car, you know, she asked me if I wanted to take her to the pig farm and stuff, where she'd liked to go out and play with the baby piglets. And I said, I just told her where to meet me if you want to go, and I went down and got my coffee and I made a big circle around. I decided right then and there if they got into my car that I was going to kill both of them, because Kara and Yolanda were inseparable, you know—one goes somewhere and the other one follows. And if they were going to get into my car that morning, that would be the last day that they lived. And Kara was the only one that went, and she got in my car."

By this point, I understood him and his motive. Yes, he was angry with Kara's mom and stepdad for throwing him out of the house. Yes, he was a habitual alcohol and drug abuser. And yes, he had had some direct conflict with Kara. But basically, he liked molesting, abusing, and raping young girls, and Kara was available to him, just as Rima had been. His entire personal history showed that nothing in his sorry life was as important or fulfilling to him as this perversity, and what made the most sense to him was to fulfill it in the easiest and most direct means available to him.

Had he in fact awakened that morning knowing he was going to exact revenge on Janet and Butch, or was it a spontaneous impulse when he saw the two girls on the sidewalk? It doesn't matter. Because Joseph Kondro didn't think it through that carefully. One way or another, he simply acted on his desire and didn't let anything stop him.

"Yolanda knew that Kara went with you, right?"

"Yeah." Which in practical terms transformed this into an even more high-risk crime than it would have been had Kondro still been associated with Kara's family, but she had been alone.

I asked what the conversation had been like in the car once he and Kara drove off.

"There wasn't really much of an exchange. I just drove her out to the house and we talked a little bit."

"But she was happy, right? She doesn't know she's in danger."

"Yeah. No, she doesn't know she's in . . . Like I said, I was playing on the . . . my victim's trust, and that's a big thing, you know. When you do these kind of crimes the person has to trust you, you know, and for you to lure them in and . . . She had no idea it will be her last day alive on this planet. And she was hungry, so I went out and I bought her breakfast at the local store out there where I committed the murder. I got some gas and took her out to the abandoned house and that's where, yeah, that's where I killed her."

Even though I'm used to this, the banality of the evil of these guys still astonishes me. It got worse:

"What went down first? What did you say to her? Like, undressed her?"

"No, she just got out of the car and ran in the house, you know, like she was playing and stuff, and I followed her in and we ended up upstairs, and I picked up this two-by-four and she was looking away from me and I just smacked her on the top of the head as hard as I could, like as if I were to have a baseball bat. And she went down, and she was on her knees and she was in a daze, you know. I hit her again. I hit her twice on the head with that two-by-four and she went out. And then I raped her, and she woke up in the middle of that act and I ended up strangling her."

"Why do you want the victims unconscious, rather than conscious?"

"It's just, it's just easier, you know. It was just easier for me."

"You don't want to hear, you don't want them to be crying, because you don't want to hear that stuff—you know, crying and begging?"

"That's just part of the act."

"So really, when [the victim is unconscious], it's almost like a masturbation, with a lifeless body, pretty much isn't it like that?"

"I don't know. If you go there . . ."

I knew exactly where I was going. "She was sexually assaulted—anally, vaginally, or what?" I already knew the answer from the medical examiner's report.

"Well, there is a controversy there because I assaulted her sexually, but the forensic pathologist said that I just anally sexually assaulted her, and that's not true. But you know you can't argue with the evidence, but I vaginally assaulted her."

You can't argue with the evidence because it's true. He did anally assault her, but as in the other case, he didn't want to admit it, because it didn't conform to his image of himself or what he wanted other people to think.

15

POWER, CONTROL, EXCITEMENT

"Joe, describe to me the feelings that you have—what they are during the sexual assault, the raping and the killing?"

As I enter into the final phase of each prison interview, I want to get the subject to summarize or affirm what was happening in his head immediately before, during, and after the crime, because the ultimate research goal since the beginning has been to correlate what was going on in the offender's mind with the evidence left at the crime scene and body dump site, the risk level of the crime both for perpetrator and for victim, and the post-offense behavior that might be observed by those around him. I particularly want to confirm the role that fantasy plays. If I had understood Kondro correctly, there wouldn't be much; indeed, far less than someone like Rader, who lived for the fantasy.

"For me, it was power, control, excitement, I guess."

That certainly squared with my impressions. "Okay—that you never felt that in your own life, or something you needed then?"

"Yeah. Like I said before, it was revenge; it was a revenge type of thing. But as far as my victims are concerned, it was just the power that I had over them, you know. You get really high when you—at least I do, I get really, really high on adrenaline—when you commit a murder like that." It wasn't the planning; that was just something he had to do. It wasn't being able to relive the assault afterward through souvenirs or his own memory; once it was done, it was done, and the aftermath was just a matter of hiding the body. Unlike most sexual predators I've encountered, Joseph Kondro seemed to live primarily in the moment.

"When you perpetrate the crime, you get this adrenaline rush, you said. But what does it feel like for you, physically, psychologically? I mean, it feels good?"

"Let me give you an example. When I was hiding Kara Rudd, that Volkswagen down there that I placed her body under, was sitting flat on the ground and I had so much adrenaline in me at that time that I actually picked that body of the car up and leaned it up against the tree by myself. And they [the police] came in and they had to use winches and all sorts of things to get that body off—I mean, to get the body of the car off her. And I just picked the thing up by myself. I am not lying when I say that. That's how much adrenaline I had going through my body at that time. It's like a high."

"So if I confront you right after a crime, well, what would you look like? How would you react?"

"I would be just like myself right now."

"You wouldn't be hyped up, sweating?" Like most sexual killers I've examined.

"No, not really. I wasn't like that when I killed my victims. I was calm as could be."

"So if I was an investigator doing an interview with you, you wouldn't—as far as nervousness, you know—be looking away from me, sweating? You think you can compose yourself in the interview?"

"I did it on the Kara Rudd case. They brought me in and interrogated me and they let me go."

This exchange told me a great deal about Kondro. Unlike Joe Mc-

Gowan, who was in a state of high excitation and agitation when he killed Joan D'Alessandro, and even when he described it to me, Kondro remained cool and in control. I would bet his pulse and blood pressure didn't even rise significantly as he spoke. And since by my examination and his own admission he was able to commit his sexual assaults and murders in the course of his normal day, he remained an extremely dangerous individual—like McGowan, but for the opposite emotional reasons.

"When we talked to you on the phone, you said it's the planning and the thinking about it that you found was satisfying to you. Can you just describe it?"

"All right. The planning part of this plays a big factor, too, because, you know, that's part of the excitement. I guess it's the fantasy. I don't actually sit down and plan this stuff—I think about it over time and then it just kind of falls into place for me."

Notice that he responded in the present tense. This kind of thinking was very much still part of his everyday existence.

"With Kara, after you were done killing her, what did you do? Where did you go?"

"Well, I wrapped her up in a blanket and put her in the front seat of my car and drove her to the Mount Solo area. I wanted to take her somewhere else, but I just wasn't sure how long it would take—because I was working against time and so I just decided to take her up on top of Mount Solo and put her there. I did; I hid her body underneath the Volkswagen."

"Why did you wrap her in a blanket?"

"Because I didn't want any blood or anything in my car, any evidence getting in my car."

"Did any evidence get in there? Did you clean your car after?"

"No, my car was pretty clean. I just wiped it down."

"When you put her in, in that Volkswagen, in retrospect, was that the right thing to do?"

"For me at the time, it was. I was just trying to get rid of the body, and that was as good a hiding place as any, you know. If I were to take her to the place that I had planned, they probably would have never found her."

"So you took her to a place that was familiar, a place you felt comfortable?" Criminals will generally gravitate to their own comfort zones.

"Yeah, it was familiar for me."

"At any point, then, because people saw you and you were attacking and killing someone you knew, you would have had to establish some type of alibi. What did you do?"

"I went and put out applications for work."

"How did you cover the time frame?"

"I tried to make sure that the people I talked to knew what it was, you know. Like I say, 'I am sorry I am late, you know,' or just tried to establish a timeline in different ways with different people. But it didn't work."

"Then where did you go after that? Did you try to go home?"

"Yeah. First I took my shoes and my shirt and stuff like that and got rid of those."

"Why?"

"Because I didn't want any footprints to be identified with my shoes. So I took them and put them in one garbage can on one side of town, took them on the other side of town, put the shoe in the other garbage can, threw the shirt out the window in an alley of an automotive business, in a mud puddle. And then I went home and changed, took a shower, washed my clothes, and then I had a parent . . . no, it wasn't a parent-teacher thing; it was I had to pick up my son and bring him to school. So I went over to Julie's house and I just picked him up."

What's fascinating to me here, even though he confuses having a parent-teacher conference with having to pick up his son, is that his mind is so compartmentalized that after raping and murdering a young girl and trying to destroy all of the evidence, he focuses on details relating to his own child with absolutely no remorse or reflection whatsoever. As a sexual predator, he knows he has to assault and rape young girls. But as a father, he knows he has to attend parent-teacher conferences and get his son to school on time.

He essentially acknowledged this attitude in our next exchange:

"Did you talk to the family at all?"

"I think I called them a couple of times. But Janet called me, accused me of taking her child and stuff, and obviously I denied it."

"Plus, this was a revenge killing, right, so . . . ?"

"At that point I didn't care what she thought or anything. I have already completed the act and that was it, you know. I just went about my business. I didn't have a second thought about it." This time, he didn't even bother to pick up on my prompt about revenge. He knew revenge was never foremost in his mind.

And though it might be difficult for most of us to comprehend, many of these guys are just as good as Kondro at compartmentalizing—raping and killing a teenage girl but getting to his son's parent-teacher conference on time.

John Wayne Gacy Jr. was a successful and outgoing building contractor in the Chicago area, with a wife and two stepdaughters. He was active in civic affairs and participated in Democratic Party politics, being famously photographed with First Lady Rosalynn Carter. Even more famously, through his membership in the local Moose Club, he performed as a clown in parades, at fundraising events, and for sick children in local hospitals. When he wasn't involved in those various activities, he found the time to rape and murder at least thirty-three young boys, whom he buried in the crawl space and backyard of his Norwood Park ranch house or submerged in the Des Plaines River.

Green River Killer Gary Ridgway, whom as an UNSUB I hunted for much of my FBI career, was married and gainfully employed. British-born serial killer David Russell Williams was a married colonel in the Canadian Armed Forces. Predators like these easily exist on two distinct psychic planes.

This is what separates the criminal sociopath from the rest of us.

"And you felt that you wouldn't get caught in this one?" I asked Kondro. "You felt comfortable?"

"I felt pretty comfortable about it. But they found her."

"It didn't scare you at all? Did you drink afterwards? If I was looking at you, would I have seen a change in your behavior, reacting differently?"

"No, I went about my life. I was as calm as I am right now."

"Really?"

"Yeah."

"You weren't immediately charged in this case. Why was that?"

"Because they didn't have a body to charge me with, you know, murder, and you have to, you have to have a body."

"Did you ever think of DNA? Did you ever think of that at any point?"

"Yeah, I thought about it, but I wasn't worried."

"Why?"

"At that point in my life, I just didn't care anymore. I was, yeah, I was out of control, you know. I would have, I would have kept doing that, you know, until they caught me." *That,* I believe.

"Do you think that chemical castration or actual castration would make a difference?"

"No." I agree with him on this point, but I followed up by asking, "Why's that?"

"These guys, they got it in their head. I think it's a genetic form that's been handed down to the generations. For me, I have thought about doing that stuff when I was a kid, you know, and throughout my whole life until the time I was caught. That's what I thought about, you know—having sex with young girls. And I wasn't going to stop."

Yes, Kondro was adopted, and it is possible he is correct that his parents regretted the act, but I could find nothing in his background or relationship with his mother or father to suggest that anything they did or failed to do contributed significantly to his evolution as a sexual predator and killer. In the nature-versus-nurture equation,

Kondro stands as evidence that some individuals are born that way and will grow up with these dangerous tendencies unless there is dramatic and timely intervention. And frankly, the data is far from conclusive as to whether that works in the majority of cases.

"So you were confident," I went on. "When they finally approached you and later arrested you, was that a shock to you?"

"No, it wasn't a shock. I was at a friend's house and they just came and knocked on the door, that I was under arrest for tampering with the witness because they talked to my ex-wife and they told me not to [talk to her about the case]. That was their excuse to keep me locked down. And I guess two other girls came forward and said that I sexually assaulted them or molested them or something, and they added those charges."

"Those other two children—did you molest those other two?"

"Yeah."

"Yeah? What were their ages?"

"I don't know. I think they were young. They were pretty young."

"Had you not been identified, would you have kept going?"

"Yeah."

"You would have kept killing?"

"Yeah."

"You felt comfortable with that?"

"I felt real comfortable about killing. I mean, killing for me, it was just like second nature now, you know. I always kept on going. It might not have been children; it could have been anybody, you know, at that time of my life, anybody that made me mad or I thought was, you know, dissing me or something."

But as we've established, his killing really had little to do with exacting revenge or punishing the victim or any other motive. It was simply something he wanted to do.

I'D ASKED HIM A LOT, AND HE'D ANSWERED ALMOST EVERY QUESTION I'D THROWN AT him. Still I had one lingering thought—the question of risk. To me, that was one of the most interesting parts of Kondro's case and thought processes.

"It's unusual in your case," I observed, "that you really knew the victims, knew the families. I mean, it was a high risk on your part and I still don't really know, don't understand, why you would go after the people you knew, versus total strangers."

"Like I said, it was the trust; it's a trust issue and more convenient for me. I tried to do this with the strangers or something like that, you know—they always fight back. You know, they tried to get away or whatever, and I just didn't want to deal with that kind of . . . you know. I wanted everything to run perfectly, and it did. I accomplished what I set out to do."

When you conduct as many interviews as I and my colleagues have, you pretty much know in advance certain things the offender is going to tell you. There are almost always similarities in the pattern. Joseph Kondro was a difficult child in a somewhat dysfunctional family. He let out his anger and frustrations by bullying children smaller and weaker than him. He developed an alcohol and drug addiction problem, which is not a causal factor for his violent behavior, but his addictions made him more impulsive—impulsive, but not careless as he planned his crimes.

The primary significant factor that makes this case important was his M.O.—an M.O. that initially fooled and sidetracked investigators. Kondro believed it was safer to target the children of his friends than to go out hunting for strangers. He didn't have to worry about control, which is a major issue and challenge with offenders who target strangers. He had already gained their trust and his targeted victims went with him willingly. Kondro had a completely different perception of what constitutes a high-risk crime.

An M.O. involving trust has been employed by offenders across the spectrum, from the most violent all the way to nonviolent criminal-enterprise-motivated predators. Ed Kemper told us that when he stopped to pick up a hitchhiking coed around Santa Cruz, he would ask her where she was headed, glance at his watch, and shake his head, as if thinking to himself whether he had time to take her there, before finally agreeing to give her a ride. In so doing, he would emotionally disarm his potential victim and she would let down her guard.

In the same way, when potential clients approached billion-dollar Ponzi schemer Bernie Madoff, he would tell them his hedge fund was fully subscribed and that he didn't need any new investors. When they persisted, since he had a reputation for returning hefty and predictable profits, he would "reluctantly" make a special exception for them, but then tell them to leave him alone with their investment and just collect the profits, because he was too busy to be involved with "small" investors.

In both cases, creating trust was the M.O., and by pretending convincingly that he really didn't want to do what he fully intended to do, the offender was able to target and exploit an innocent victim.

Both friends and police overlooked Joseph Kondro as a suspect because he feigned empathy and concern for the missing children and even participated in searches to find them. We have seen that behavior before in stranger abductions, where an UNSUB will participate in searches and inject himself into the investigations. We even see it in semi-stranger scenarios, as we did with Joseph McGowan. Oftentimes the offender will try to provide bogus information to steer the investigation away from him.

In summary, Kondro's background was typical, but his victim preference—his friends' children—was something I had never seen in a case until this one.

Lesson learned: Everyone is a potential suspect, and don't let looks or behavior fool you.

Note the juxtaposition of what we refer to as the signature and modus operandi elements of the crime: He is coldly dispassionate and analytical in the way he describes determining how to carry out his perverse psychopathology.

And like a great many other killers and predators, Joseph Kondro came to a sense of his heritage and spirituality—or claimed to—only *after* he was in jail, with no immediate thrills or stimulation in his life. Often violent predators find spirituality in prison, or at least claim to. "I'm a Christian, you know," Dennis Rader told me. "Always have been. After I killed the Oteros [five members of one family, his first known victims], I began to pray to God for help so I could fight this thing inside me. My greatest fear, even more than being caught, was whether God would allow me into heaven or would I be condemned forever." I didn't doubt his sincere belief in the afterlife. In fact, I thought this was pretty interesting, since I knew he'd held the belief that the women he killed would serve him as sex slaves in the next world. The fact that he could even contemplate the idea that he would be allowed into heaven after what he'd done speaks to his malignant narcissism.

"As part of the spiritual, you know—that higher life—are you supposed to get things off your chest?" I asked Kondro. "Are you supposed to admit to all your sins, any of your sins, before you can go to that higher form of life?"

"It's always good to clear yourself of wrongdoing before you—before you pass on. Yeah, somebody once told me that, you know, and I still believe it. You know, you have to, like you said, get rid of your sins or your wrongdoings, or you walk endlessly through the spiritual world, you know; you won't cross over. You will just end up being a lost spirit."

I have real trouble believing that a man who is so casual and cavalier about raping and murdering children could be that concerned with being a lost spirit.

"You don't want to be a lost spirit?"

"No. I think I was a lost spirit when I was born." Again, nothing is really his fault.

"But you can—you can change things around? Is that your spiritual belief now?"

"I don't know about changing things. That's up to the creator."

"Any regrets?"

"No. I have a regret that I should have been a better father to my children. But I have no regrets, no."

"No regrets, back to 1985 with your first victim? Any regrets there?"

"No."

"How about 1996? Any regrets there?"

"No."

"Anything you would have done differently?"

"Yeah, I would have took her [Kara Rudd] out to a different spot."

"So she wouldn't have been found?"

"Well, that was the idea. That was the intention."

"Do you think you are crazy?"

"Yeah, I think I have some issues."

"You know, people in the past referred to you as being a monster. What do you think? You think you are a monster?"

"Before I came to prison I was. Now I am just institutionalized."

"Are you receiving any kind of treatment here?"

"No."

"Could you, if you want to, go to counseling?"

"Yeah. I see a mental health counselor once in a while, but he just . . . they got me on some drugs that help me sleep at night. But I really have never sat down like I am sitting down with you and discuss my case with anybody, you know."

"Did you ever want to, or just not interested?"

"No, I have never wanted to. Never until—until now."

"Why did you decide to agree on the interview?"

"There has been a lot of people in the media world that have contacted me, asking me for interviews, and I have always denied them. This time of my life I feel that there should be closure, not just for the families but for me, you know. It's a big step in my life to put this behind me."

DUE TO THE DESCRIPTIONS OF EXPLICIT VIOLENCE IN OUR INTERVIEW, THE DOCUMENtary could present only a small portion of what has been recounted here, for the first time publicly. Ironically, when the executives at MSNBC saw the interview tapes, they were much more impressed with Joseph Kondro's "performance" than with mine. They wanted me to be confrontational—to get in his face and register outrage over the horrible things he had done. I think they wanted me to be supercop.

Whatever my personal feelings, what is important to me in all prison interviews with violent offenders is not to express my own moral outrage, which accomplishes nothing, but to learn as much as I can by establishing a connection with the subject. In fact, when I read Linda Stasi's review of the program in the *New York Post*, I was surprised at her observation about one of my reactions: "When the normally unflappable Douglas finally gets disgusted enough about Kondro murdering his friend's little eight-year-old daughter, he says to him, 'But you killed a little girl!' " This was probably the most I've let down my guard in one of these situations.

While it's tempting to play a real-life Clint Eastwood character before the cameras, and it certainly would be emotionally satisfying to beat the crap out of him (though in real life they would put you in jail for this), I tried to explain that any approach other than what I had always used would be counterproductive, that this was how we had done it effectively when I was in the bureau. My role is to get these guys to talk, to find out what is, and was, going on inside their minds. Confrontation and moral indignation do not achieve that. In the end, talking to killers is about playing the long game, with every move a deliberate one—outrage, anger, these emotions are ever present in the background, but they work against you only if they come to the surface.

Though the reviews were gratifying, that apparently was not what MSNBC was looking for, and the series was not picked up. In fact, it may be that the only way to properly convey what a deep dive into the killer's psyche is like is in a book like this, though the Netflix *Mind-*

hunter dramatic series has certainly conveyed the mood and feeling of these mental confrontations.

On May 3, 2012, Joseph Robert Kondro passed away at the prison in Walla Walla, aged fifty-two. His brief obituary stated that he died of natural causes. The death certificate listed the specific reason as end-stage liver disease due to hepatitis C. He was buried in the Pinery Indian Cemetery in Baraga County, Michigan.

In the local newspaper, Kara's mom Janet was quoted as calling the death "a great weight lifted off."

Kondro was unique in our studies and experience, the only repeat rapist-murderer who targeted the children of people he knew well. That in itself added to our store of knowledge and understanding. It opened our eyes to the idea that anyone can be a suspect and that the traditional indicators of innocence—cooperation with the police, participating in search parties, offering alibis, and all of the others—have to be evaluated in context and against the circumstantial evidence.

Kondro possessed addictive behavior traits. He relied on drugs and alcohol. And in his case, his addictive behavior spilled over into sexual molestation, rape, and homicide. I learned from the Kondro interview that when we are analyzing a series of cases—any types of cases—the offender's underlying psychological needs may be power and control, but they are connected directly to a lack of impulse control on the part of the offender. That is related to the addictive personality. This does not, however, mean that he cannot control his actions. In Kondro's case, the impulse to power and control through sexual attack went hand in hand with careful planning before he committed the offense and careful cover-up and establishment of his alibi afterward. Alcohol and drugs may cause the offender to be careless and take risks, but he doesn't want to get caught. The "death wish" is only for his victims, not for himself.

What does this all mean to an investigator? First, we have to scrutinize our own instincts and biases that no one close to the victim who doesn't have an obvious motive would do such a thing and therefore can be ruled out as a suspect. Kondro had been brought in on several occasions for questioning. As Hamlet said to his best friend, "There are more things in heaven and earth, Horatio, than are dreamt of in your philosophy."

Second, if it is determined through investigative means that a certain individual in some way close to or connected with the victim appears to have an addictive personality and/or questionable impulse control, that individual should be placed high up on the suspect list.

Third, Kondro reaffirms for us that a casually violent personality—one capable of ripping a phone out of the wall, breaking household objects in a fit of rage, or leading a domestic partner to seek a restraining order—is by definition capable of a heightened or intensified level of violence. Because as we say, "Behavior reflects personality."

And that awareness just might save a life.

III
ANGEL
OF DEATH

16

PLAYING GOD

A lot of people are surprised to discover that the most successful and numerically devastating serial killers are not the Jack the Rippers, who stalk their victims by night, or even the Ted Bundys, who charm, entice, then abduct, assault, and kill beautiful young women. Actually, and ironically, they are workers in one of the most sacred helping professions, people who don't even bother to hide their identities or their faces from unsuspecting victims or their families. And in many cases, it takes years before authorities even realize a crime has been committed, so the case is cold.

Donald Harvey, a mild-mannered man from Ohio, with a ready smile, a sunny disposition, and an endearing manner, may be the most prolific serial killer in American history. Between 1970 and 1987, he may have killed as many as eighty-seven people, all while hiding in plain sight, just like Joseph Kondro. But he was as different from Kondro as Kondro was from Joseph McGowan. His preferential victims: elderly men and women in hospitals who could not resist or fight back. By the time he was captured and brought to justice, he had proudly embraced the moniker Angel of Death.

He was the second killer I confronted for the MSNBC pilot.

When Harvey was finally apprehended and interrogated, I was

asked to observe the interrogation and provide on-site consultation as to how to get him to confess. As it turned out, though, Harvey spoke freely to the FBI agents questioning him. The interrogators were interested in his motivation so they could help put together a coherent case for trial. I was interested not just in his motivation but also in his actual behavior before, during, and after each crime. How and why did Harvey become the person he was? How did he learn to kill so efficiently, and what factors went into his going undetected for so long? Was he born with homicidal impulses or were they the result of a dysfunctional upbringing that demanded some form of displaced retaliation or retribution? What could have been done to stop him and what can now be done to keep crimes such as his from happening again? This was what I hoped to get out of the interview.

If Donald Harvey was a one-off type of predator, he would be interesting and grotesque, but not necessarily significant from an investigative perspective. Unfortunately, he is the prime propagator of a type of murder that has its own category in the *Crime Classification Manual*: Medical Murder.

Every serial killer is horrible and terrifying in his own unique way. But on a certain level, what Donald Harvey did was particularly despicable because, like Kondro, he played on trust. But unlike Kondro, he perverted one of our most cherished values: the mission to heal and comfort the sick.

These may not be the types of criminals that thriller movies are made about. We may not think of the Donald Harveys as being dangerous in the same way we would a Ted Bundy, a Charles Manson, or a John Wayne Gacy. But that would be a mistake. Most of us, fortunately, will never encounter that type. But just because the Harveys don't lurk in the shadows, waiting to pounce on unsuspecting victims going about their daily lives, doesn't mean they are not as dangerous. Their victims are often the same as the more traditional serial killers: the elderly and defenseless. We think of young, beautiful women as the targets of these men, but vulnerability, more than anything else, is likely to lead to someone's becoming a victim. That's why children, senior citizens, prostitutes, drug addicts, the homeless, and other marginalized groups are prime targets of serial killers.

Individually, as we've said, the Harvey kind of predator can be a lot more "productive" as serial killers. Almost all of us at some point will find ourselves in a health-care setting, either for ourselves or a family member.

For most of us, hospitals are pretty scary places to begin with. Even in the case of "routine" procedures, we can never be certain of the outcome. If we get to the point where we feel we can't trust the man or woman behind the surgical mask or the nurse's badge, then the hospital becomes the stuff of nightmares or horror movies.

When I think of times when I've been hospitalized, it never occurred to me that a cross word uttered in pain or frustration to a nurse could result in that person's deciding to kill me. Or when I was comatose in Swedish Hospital in Seattle after suffering viral encephalitis while hunting the Green River Killer in December 1983, what if some nurse or orderly playing God decided to "put me out of my misery"? Anyone could be the next victim.

Since police and detectives are not even generally aware or on the lookout for this type of criminal, for me the most important behavioral consideration for an interview like this is: *How do we recognize and identify these killers?*

DONALD HARVEY WAS BORN APRIL 15, 1952, IN BUTLER COUNTY, OHIO, TO RAY AND Goldie Harvey, the first of their three children. Shortly after his birth, his parents moved to Booneville, Kentucky, a remote rural community on the eastern slope of the Cumberland Mountains, in the Appalachian chain. Ray and Goldie were tobacco farmers. They were religiously observant and became regular attendees at the local Baptist church. By all accounts, Donald was a good and good-looking boy, with dark curly hair and big brown eyes, who didn't give anyone any trouble. The principal of his elementary school recalled him as happy, sociable, well dressed and groomed, and popular with the other children. Some former classmates remembered him being something of a loner who played up to the teachers.

He was, reportedly, a good student at Booneville High School, earning mostly As and Bs, but became bored with studies and dropped out before graduation. He took some correspondence courses while working in a store selling tennis and golf equipment and earned a GED—General Educational Development—equivalency certification by the time he was sixteen. All in all, a fairly prosaic story.

Yet this seemingly average childhood masked several elements that, individually or collectively, may have had a significant influence on Harvey's future. When Donald was six months old, his father fell asleep while holding him and the baby dropped to the floor. He didn't appear to be seriously injured. When he was five, he fell off the running board of a truck and struck his head. He never lost consciousness, but he sustained a five-inch-long cut on the back of his head. And throughout his childhood, according to various reports, Donald's parents had a tense, sometimes abusive relationship with each other, though Goldie stated that her son was brought up in a loving family.

Did these traumas contribute to the emotional fashioning of the man Donald Harvey became? There is an ongoing debate in the forensic medical community about the influence of brain injuries and abnormalities on the commission of violent crime. A number of killers, subjected to imaging studies or as a result of postmortem examination, have been found to have brain lesions of various sorts. Those in the deterministic camp, who believe that much aberrant behavior is influenced by distinct physiological causes, point to these lesions as proof that this is why the criminal acted the way he did. Those in the "free will" camp suggest that these lesions may be more symptom than cause—that is, they are the result of injuries produced by the impulsive, risk-taking behavior that these guys display as children.

In Donald Harvey's case, there is no evidence that these accidents had any effect, but other events in his childhood offer reasons for concern. Because serial predatory offenders often come from dysfunctional families or have backgrounds that include abuse, I was especially interested to find out during my preparation for the interview that there were incidents that almost certainly had some effect on Harvey's developing psyche. From around the age of four on, according to a detailed report developed by the Department of Psychology at Radford University in Virginia, Harvey was sexually abused by his mother's half brother Wayne, who forced him into oral sex and used him as an aid to masturbation. Perhaps a year later, young Donald was also subjected to sexual overtures from an older neighbor. Both individuals were able to maintain their relationships with him until he was about twenty.

Children can easily be threatened and manipulated into not telling anyone about the abuse. But when children psychically strong enough reach a certain age, they will resist or tell someone. Harvey got to the point where he was old enough to say no to these sexual advances, yet he did not. By that point, he had figured out that he could manipulate his uncle and the neighbor, extorting them both to get what he wanted. He later observed that he liked it when the neighbor gave him money. Finally, around the time he received his GED, at age sixteen, he had his first consensual sexual encounter. The next year, he began an on-and-off relationship with another man that continued for about fifteen years.

Bored with his hometown, Harvey moved to Cincinnati, Ohio, and found a job at a local factory. But work at the factory slowed and he was laid off. A few days later, his mother called and asked him to go visit his grandfather, who was in Marymount Hospital, a Catholic institution in London, Kentucky, not far from Booneville. Being unemployed, Harvey willingly agreed to go.

Harvey spent a lot of time at the hospital and quickly ingratiated himself with the nuns who served as nurses and administrators. This natural ability to charm and befriend others in a professional environment was a trait he would employ again and again. He just seemed like the kind of young man who was eager to please and always had everyone's best interests at heart. One of the nuns asked him if he'd like to work at the hospital. He needed a job and didn't particularly want to go back to a factory, so he happily accepted. He didn't have any hospital or health-care experience, but he became an orderly, responsible for cleaning up patients and their beds, changing bedpans, moving patients around the hospital for tests, inserting catheters, and dispensing medication, among other duties. Harvey enjoyed the work and the hospital staff liked him and appreciated his willing attitude.

During an evening shift on May 30, 1970, the eighteen-year-old orderly went in to check on Logan Evans, an eighty-eight-year-old stroke victim. The nun who was attending him told Harvey that he had pulled out his IV tube and would need to be cleaned up and have the line reinserted.

When Harvey pulled back the bedsheet, he saw that Evans had defecated on his own hand. As he leaned in, the patient rubbed the soiled sheet in his face. Harvey became enraged and spontaneously smothered Evans with a pillow wrapped in blue plastic sheeting. "It was like the last straw," he later recalled. "I just lost it. I went in to help the man and he wants to rub that in my face."

As Harvey continued to hold the pillow over Evans's nose and mouth, he listened to his fading heartbeat with a stethoscope until it ceased. Then he disposed of the plastic sheet, cleaned up the corpse, changed the bed linen, dressed Evans in a fresh hospital gown, and took a shower. He went out and informed the nurse on duty that Mr. Evans had apparently died.

The deceased patient was taken to a funeral home and as Harvey commented, "No one ever questioned it."

Regardless of the particulars of the murder, from a law enforcement standpoint, this is about the worst thing that can happen, because it can pave the way for a career as a serial killer. Once the new offender realizes he has gotten away with it, his sense of personal power is enhanced and he begins to create his own mythology—that is, he is smarter than the police and everyone else around him.

Dennis Rader and David Berkowitz were certainly examples of this. Even though neither one was exactly the brightest bulb in the pack, both saw the fact that they got away with a series of murders as evidence of their own intelligence and law enforcement stupidity. As is true with ordinary people, serial predators can mistake luck for personal ability. And we have to admit that luck often plays a critical role in how quickly an UNSUB is identified or an offender is apprehended.

When Berkowitz got away with several murders, he started believing he was a master killer. He was getting full press treatment and there was a hundred-member task force devoted to finding him. With all that, he reasoned, he must be pretty good at this. He wrote a semicoherent letter to the NYPD's lead detective on the case, Captain Joseph Borelli, who later became chief of detectives. In it he gave himself the name Son of Sam, addressed himself like a monarch to "the people of Queens," and signed himself, "Yours in murder / Mr. Monster."

What Berkowitz did not understand was that just as luck could protect him, it could also run out, as happened when a parking ticket connected his name to the last of the crime scenes.

In certain ways, Berkowitz and Harvey present similar motivational models. Though Berkowitz was heterosexual and Harvey was gay, they were both sexually stunted in their formative years. Harvey was molested as a child by people he knew and should have been able to trust. Berkowitz's first sexual encounter was with a prostitute while he was serving in the army, and he contracted a venereal disease from her. Based on their life experiences and personal skills, the two men would kill differently—Berkowitz with a powerful handgun and Harvey with health-care equipment and drugs—but both would kill out of resentment and low self-esteem.

One of the most interesting insights that came out of the Berkowitz interview was how he was always thinking about these murders, to the point that on a night when no victims of opportunity were available, he would return to the sites where he had successfully killed to masturbate and relive the sensation of power and sexual energy he derived from the crime itself.

Contrast that with Donald Harvey, who had a nearly endless supply of opportunistic victims available to him and didn't have to go on the hunt at all. Harvey, who actually was pretty intelligent compared to someone like Berkowitz, realized that by carefully observing his surroundings and using the workings of the system—in this case, the hospital routine—against itself, he could pretty much do what he liked. He bragged to me that he could quickly assess any hospital's security weaknesses. For example, he thought it would have been more difficult for him to kill had management rotated his shift more often or regularly assigned him to different departments throughout the hospital. The fact that he was assigned to a single ward, with the same staff and patient population, gave him the confidence and comfort to go about his nefarious business without being detected.

It wasn't long before this realization was confirmed for him.

The next day, he used the wrong-size catheter on James Tyree. Harvey claimed this was accidental, but when Tyree yelled for him to remove it, Harvey restrained him with the heel of his hand until the patient vomited blood and died.

Just three weeks later, Harvey was in the room of an elderly woman named Elizabeth Wyatt, who told him she was praying for death as a relief from her suffering and wished she could simply die of her own will. Harvey obliged by turning down her oxygen supply. Several hours later, a nurse found her dead.

The following month, he turned Eugene McQueen on his stomach so that the respiration-impaired patient could not breathe. He drowned in his own fluids, after which Harvey followed a nurse's order to bathe him. When McQueen was pronounced dead, rather than investigate the death, several members of the staff teased Harvey for bathing the patient, not knowing he was dead.

Less than two months after killing Logan Evans in his first murder, Harvey came in to catheterize Ben Gilbert. But the patient knocked him out with a metal urinal. The disoriented or mentally unsettled Gilbert apparently thought Harvey was a burglar who had come to rob him. When Harvey recovered his senses, he decided to seek revenge. That evening he went back to Gilbert's room, and instead of fitting him with the appropriate 18 Fr catheter, he employed a larger diameter, 20 Fr, used for women. He then unbent a wire coat hanger and fed it through the catheter, puncturing the bladder and intestines. Gilbert went into shock from internal bleeding and then sank into a coma. Harvey removed the wire and catheter and disposed of them. Then he refitted an 18 Fr catheter and reported that he had come in to find the patient unresponsive. Gilbert died of fulminant infection four days later. This was the first murder that we can definitively say was premeditated.

As improbable as it seems, no one appears to have speculated about any of these deaths or correlated Harvey's presence in the specific hospital rooms. Each time he got away with it, his attitude of superiority and resourcefulness was strengthened.

Before he reached his nineteenth birthday, Harvey had killed at least fifteen patients at Marymount.

Each time he leveraged the patient's and the system's vulnerabilities to his advantage. He used a faulty oxygen tank with Harvey Williams and Maude Nichols. He later said the Williams death was an accident, but he killed Ms. Nichols because she had been brought in with bedsores so infected that they were maggot-infested, and staff members were reluctant to give her proper care. He simply did not turn on William Bowling's oxygen because the patient was having to struggle so hard to breathe. He then died of a massive heart attack. Viola Reed Wyan was suffering from leukemia and Harvey thought she smelled bad. He decided to end her suffering as he had killed Logan Evans—with a pillow wrapped in a plastic sheet. But someone came in and he had to stop. So later he hooked her up to a faulty oxygen tank and waited for her to die.

None of these deaths was investigated for anything other than what it appeared at first glance, so Harvey was able to continue adding victims. It was clear to me that Harvey saw himself as an underachiever. He was an orderly in an environment where the doctors and nurses got all the status and respect. He was going to prove that he was better at the "game" than they were. In fact, he was going to make a mockery of it.

Several times he tried to smother Silas Butner, who was suffering from kidney failure, but each time he was interrupted. Harvey employed the faulty oxygen tank again. Likewise, he decided that Sam Carroll had suffered enough from pneumonia and an intestinal blockage, and John Combs had put up with his heart failure long enough. He killed them both with a malfunctioning oxygen tank. He was able to smother Maggie Rawlings, who was being treated for a burn on her arm. He put a plastic bag between her face and the pillow so that if anyone checked her postmortem, they would not discover any fibers in her airway.

He also employed painkillers. He killed Margaret Harrison by giving her an overdose of Demerol, morphine, and codeine, all of which had been intended for another patient. He killed Milton Bryant Sasser, who had been admitted with congestive heart failure, with an overdose of morphine stolen from the medicine locker at the nurse's station. When he tried to flush the hypodermic needle down the toilet, it got stuck and clogged the pipe. But no one made a connection with Sasser's death.

As if this trail of death from his first year at the hospital weren't enough, Harvey additionally began to learn more about how dead bodies present. During that year, Harvey began a relationship with Vernon Midden, an undertaker who was married with children. Midden taught him a lot about dead bodies and the physical evidence of how they died. Harvey was particularly interested in how to hide or mask the indications of smothering and asphyxiation.

It was information he would put to use in the near future.

TOWARD THE END OF MARCH 1971, HE LEFT HIS JOB AT MARYMOUNT HOSPITAL, POSsibly because he feared that the number of his crimes might be catching up with him. It is also possible he was simply depressed, because that spring he set fire to the bathroom of an empty apartment in the building where he lived, figuring to commit suicide through asphyxiation. But he wasn't nearly as proficient at killing himself as he had been with others. He was arrested for destruction of property and paid a $50 fine.

Shortly thereafter, he was arrested for burglary. In a drunken state, he babbled to the arresting officers about having killed fifteen people at Marymount. The officers questioned him about these statements and tried to check out his claims, but there was no evidence and no one at the hospital believed he had anything to do with the deaths. He pled guilty to a reduced charge of petty theft and paid another small fine.

During this time, according to the Radford University report, Harvey had his first heterosexual encounter, with a woman named Ruth Anne Hodges, with whose family he was staying while applying for a job in Frankfort, Kentucky. He recalled being naked with her during a drunken evening but denied remembering anything else. Nine months after this, Hodges gave birth to a son, naming Harvey as the father. Over the years, he alternately accepted and denied paternity.

In June 1971, Harvey enlisted in the U.S. Air Force, during which time he had a brief affair with a man named Jim, whom he later admitted he had the urge to kill. He apparently was deterred by fear of being caught while in such a highly regulated setting as the military.

But he didn't last long in the service. He made another suicide attempt, this time with an overdose of NyQuil. This led Air Force authorities to learn about his arrest and his apparently mad ravings to the police in Kentucky about killing people in a hospital. The Air Force didn't want any repeats, so he was given a general discharge in March 1972.

After continued depression and an argument with his family, he tried to kill himself again, this time with an overdose of the sedative Placidyl and the tranquilizer Equanil. He was taken to the hospital, where they pumped his stomach and then transferred him to the mental ward at the Veterans Administration Medical Center in Lexington, Kentucky, where he was often in restraints because he couldn't control himself. He was given repeated electroshock therapy, which is often highly effective in treating chronic depression. But in Donald Harvey's case, it had little effect. When he was released, his parents informed him he was no longer welcome in their house.

Over the next several months in 1972, Harvey worked part-time as a nurse's aide at Cardinal Hill Rehabilitation Hospital in Lexington, while continuing to receive outpatient therapy at the nearby VA hospital. He took a second job at Good Samaritan Hospital. During this time, he had two relationships with men, each of whom he lived with intermittently.

But the most significant aspect of this time was that at neither hospital did Harvey try to kill anyone. This might have been because he was trying to control his urges and compulsion to play God. Or it could have been that he was more tightly supervised than he had been at Marymount and feared he might be caught. The hospital environment was his zone of comfort, and if he couldn't commit crimes there with his victims of preference, it would be even more daunting to commit them elsewhere with others.

But things were about to change again.

17

WORKING NIGHTS

In September 1975, Harvey moved back to Cincinnati and landed a job working nights at the VA Medical Center there. He had a wide range of duties and filled in wherever he was needed—as a nurse's aide, cardiac catheterization technician, housekeeper, and morgue assistant. With this variety of responsibilities, he had virtually unlimited unsupervised access to all areas of the hospital. He particularly liked working in the morgue and tried to learn all he could.

It was in this role that Harvey's interest in the occult blossomed. For years he had been fascinated by the magic and mystical-related, but his interests hadn't found a proper home until now. He wanted to join a group dedicated to subjects relating to occultism and the dark arts. But there was a problem: This group accepted only heterosexual couples as initiates and he was a single gay man. So Harvey and another man were paired with the wives or female partners of men already in the group for the performance of the initiation rites.

There has been much discussion over the years about the relationship between Satanism, the occult, witchcraft, black magic—or anything else we may call it—and violent crime. For a while in the 1980s and 1990s, it was a subject on nearly every television talk show, and police departments actually hired consultants to teach them how to recognize homicides that were satanic ritual murders. Though the media, the public, and much of law enforcement was convinced these ritualistic crimes were happening all over, the FBI looked intensively into each claim and didn't find a single legitimate example. My friend and highly respected colleague, former Special Agent Kenneth Lanning, wrote a groundbreaking treatise around this time, entitled *Investigator's Guide to Allegations of "Ritual" Child Abuse*, in which he basically debunked the entire phenomenon. Likening the claims to the numerous accounts of alien abduction, Ken wrote, "In none of the cases of which I am aware has any evidence of a well-organized satanic cult been found."

But how does that square with the fact that an active serial killer like Donald Harvey had a profound interest in the occult and joined a group devoted to it?

The basic answer is that Harvey's kills were not motivated by the occult, nor did they have any ritualistic or quasi-religious elements to them. He may have become interested in occultism for the same reasons that he killed—dissatisfaction with himself and a yearning for power—but it wasn't Satanism or black magic that made him kill. Let's call it a symptom rather than a cause. So what we are left with is an inadequate personality who is also trying to gain power or another satisfying dimension to his life through the occult, but that had nothing to do with *why* he killed. And after his relatively brief murder-free sabbatical in the Air Force and in Lexington, Harvey was soon back at it.

Over the next ten years, Donald Harvey killed at least fifteen more patients at the VA hospital, expanding his knowledge, his methods, and his ingenuity. He still cut off oxygen supplies, but he also favored suffocation, arsenic, cyanide, and old-fashioned rat poison in a patient's dessert. Cyanide has legitimate medical uses, including rapidly bringing down blood pressure, dilating blood vessels, and testing ketone levels in the blood of diabetics. Harvey found that cyanide worked equally well when it was introduced into an intravenous line or injected directly into a patient's buttock. He slowly appropriated small amounts from hospital stocks until he had amassed around thirty pounds of it at home! He also studied medical journals to refine his understanding of how to conceal his crimes. The greatest aid he had going for him, however, was that unlike other serial killers, his murders were all presumed to be natural deaths.

He was dating a man for a while, but they argued frequently. After one particularly heated dispute, Harvey slipped arsenic in the guy's ice cream. He wasn't necessarily trying to commit murder here, only make his partner sick and uncomfortable. But this act was significant in that it was the first time Harvey had tried to harm someone outside a hospital setting.

Any time something like this happens, it represents a dangerous escalation and crucial tipping point in Harvey's behavior and the risk he posed to society at large. Virtually every serial killer will begin in what we call his primary comfort zone. It may be an area close to his home, his place of work, a park that he knows well, or any other site where he feels comfortable and confident. When we get a serial killer or rapist case, we always pay special attention to the location of the first crime in the series, because it will tell us a lot about him.

Even with a killer as crazed and unstable as we profiled Jack the Ripper to be, we could discern distinctive primary and secondary comfort zones when we studied the map of London's East End and pinpointed his crime scenes in order. Harvey's primary comfort zone, clearly, was the hospital ward. No one bothered him or paid particular attention to his actions. From experience, he knew he could function freely in that setting.

Once he was prepared to attempt murder outside that comfort zone, he had progressed significantly in his evolution as a serial killer. He had evaluated the new risk factors and concluded he was prepared to accept them. Murder was now not just something he did when he found himself in a particular situation. Now the killing was what defined him, rather than the setting of the deaths. As Harvey's comfort zone expanded, there would be no safe space for his potential victims.

That same year—1980—he began a live-in relationship with another man, Carl Hoeweler. But when Harvey found out that Carl had a habit of picking up other men in a local park on Mondays, he started putting small doses of arsenic into his food on Sundays so he wouldn't be up to his Monday adventures.

Carl was friendly with a neighbor who Donald felt was threatening the relationship and trying to split them up. He tried to poison her with acrylic acid, and when that didn't work, he tried to infect her with live AIDS virus he had obtained from the hospital. But neither of those methods proved effective, so he spiked her drink with a hepatitis B serum he had also stolen from the hospital. The infection was so severe she had to be hospitalized, at which point she was properly diagnosed and treated. But still, no one connected her illness to foul play or Donald Harvey.

Another neighbor, Helen Metzger, was perceived by Harvey as a threat to Carl. He sprinkled arsenic on some leftovers he gave her and in a jar of mayonnaise in her refrigerator. Several weeks later he gave her a pie laced with more arsenic. She developed paralysis and needed a tracheotomy to facilitate breathing. But following the tracheotomy, she began hemorrhaging and lost consciousness. She never came out of it. The hospital attributed her death to Guillain-Barré syndrome, a paralyzing condition that can affect the lungs.

Harvey volunteered to be a pallbearer at his dear neighbor's funeral, and later claimed he didn't mean to give Ms. Metzger a lethal dose of arsenic, only enough to make her sick. Which is exactly what happened to members of Metzger's family gathered at her apartment following the funeral. Some of them became ill after eating mayonnaise from the sabotaged jar. Fortunately, they all recovered. Their sickness was chalked up to accidental food poisoning.

Harvey's relationship with Carl was inspiring Harvey to new heights of homicidal lust. He had an argument with Carl's parents, so he began poisoning their food with arsenic. Carl's father, Henry Hoeweler, had a stroke and was admitted to Providence Hospital. Harvey went to visit him a few days later, and while there, added more arsenic to his dessert pudding. Henry died later that night of kidney failure and the effects of the stroke. Over the next year, Harvey continued intermittently poisoning Carl's mother, Margaret, but couldn't manage to kill her.

But he did manage to kill Carl's brother-in-law, Howard Vetter, by accident. He had been removing adhesive labels with methanol, or wood alcohol, and was storing it in a vodka bottle. Carl either didn't know or confused the bottle and poured Howard several drinks. Methanol is highly toxic, and Howard almost immediately fell ill. He suffered a heart attack and died.

By January 1984, Carl had had enough of Harvey's erratic behavior and mood swings and asked him to move out. Harvey was so angry at this rejection that he made several attempts to poison Carl over the next two years, none of them successful. He did poison another ex-boyfriend, James Peluso, who had heart disease and had asked Donald to "help him out" if he got to the point where he couldn't take care of himself. Donald spiked his daiquiri with arsenic and put some in his pudding. He was brought to the VA hospital, where he died. Because of his history of cardiac problems, no autopsy was performed.

Arsenic also went into the Pepto-Bismol solution that neighbor Edward Wilson drank. Wilson and Carl had gotten into an argument over utility bills, and even though Donald had his own problems with Carl, he wanted to protect him. Wilson died five days after the poisoning.

Around the same time, the ever-diligent Harvey was promoted to morgue supervisor at the hospital, which roughly coincided with him joining the National Socialist Party, a neo-Nazi fringe group. He claimed he did not sympathize with the Nazi goals but was doing reconnaissance inside in an attempt to bring them down. When I learned this, I questioned his professed motivation. An association with an organization with an evil and hate-filled mission seemed similar to me to Harvey's fascination with occultism. One way or another,

I felt, it was all about power and the sexual charge he got out of wielding it, even if no one else knew.

Karma finally started to catch up with Donald Harvey on July 18, 1985. As he left the hospital, security guards thought he was acting suspiciously and confronted him. They demanded to search the gym bag he was carrying. Inside, they discovered a .38-caliber handgun, which was strictly against VA hospital policy. They also found several hypodermic needles, surgical gloves and scissors, and some drug paraphernalia, including a cocaine spoon. Then there were several medical texts, books on the occult, and a biography of Charles Sobhraj, the Indian-Vietnamese serial killer who preyed on Western tourists throughout Southeast Asia in the 1970s. When hospital personnel then searched Harvey's locker, there was a small liver specimen mounted in paraffin ready to be sliced and prepared for microscopic examination. Harvey claimed the gun was planted in his bag, possibly by Carl Hoeweler.

There were some errors and irregularities in the investigation, so authorities might have had a difficult time prosecuting Harvey. Moreover, they apparently didn't want any adverse publicity, so they agreed to let him resign quietly and pay a $50 fine for carrying a firearm on a federal reservation. Nothing about the incident was placed in any criminal record and no attempt was made to see if Harvey had committed any other offenses.

Seven months later, in February 1986, he found another hospital job—this time as a part-time nurse's aide at Daniel Drake Memorial in Cincinnati. From what subsequent investigations were able to determine, no one questioned Harvey about his previous jobs or why he had left them. Whatever else you say about this guy, he was certainly persistent and dedicated to his "profession."

His first kill at Drake was a semicomatose man named Nathaniel Watson, whom Harvey smothered with a wet plastic garbage can liner. He'd tried to kill Watson on several previous occasions but had been interrupted each time—something he'd experienced before and was now prepared for. His motive was curiously bifurcated this time: He didn't think Watson should have to undergo the indignity of a vegetative state in which he was being fed through a gastric tube, and he'd heard the patient was a convicted rapist—not substantiated—who deserved to die. Watson was found dead by a nurse less than an hour later.

Four days later he killed another patient, Leon Nelson, in the same manner.

He poisoned two others before receiving an employee evaluation that rated him "Good" in six out of ten categories and "Acceptable" in the other four.

Over the next ten months, Donald Harvey dispatched at least twenty-one more patients. His favorite methods had become arsenic and cyanide poisoning, though he employed Detachol, an adhesive remover used with colostomy bags, through gastric tubes in two patients, one male and one female.

During this time span, Harvey was having his own personal problems. His relationship with Carl finally broke up for good and he started seeing a therapist for depression. He became more obsessed by occult rituals, and he once again attempted to commit suicide, this time by driving his car off a mountain road. He survived but came away with head injuries, returning to work at the hospital and his eventual undoing.

JOHN POWELL, FORTY-FOUR, HAD BEEN COMATOSE FOR SEVERAL MONTHS AS THE RE-sult of a motorcycle accident in which he had not been wearing a helmet. He had shown some subtle signs of recovery but was in general decline and not much hope was held out for his recovery. Doctors, therefore, were not particularly surprised when he died suddenly. It was the policy of the Hamilton County coroner's office that all motor vehicle deaths receive autopsies, so one was ordered to see if the exact cause of death could be ascertained.

The postmortem examination was performed by Dr. Lee Lehman,

who was both a forensic psychologist and an expert in biochemistry. As soon as he opened the body cavity, he noted a telltale odor, which some have likened to the scent of bitter almonds. Dr. Lehman knew immediately to associate this smell with cyanide, and murder went to the top of his differential diagnosis as to cause of death.

"I don't know what bitter almonds smell like, but I know what cyanide smells like," Dr. Lehman told Howard Wilkinson of the *Cincinnati Enquirer.*

He completed the autopsy and prepared tissue samples to send to three other labs to confirm his suspicion.

All three laboratory reports came back positive for cyanide.

Lehman notified the Cincinnati police department, and a criminal investigation was launched into John Powell's death. Detectives logically focused on Powell's family and friends and others with whom he had had contact, beginning with his wife, who was brought in for interrogation. This is a standard procedure—perhaps with her husband in a vegetative state and bills piling up, she wanted the entire ordeal to be over. But they could find no motive or any evidence that she or anyone else in the family wanted him dead or bore him any ill will.

The next logical step was to look at hospital personnel who had had access to Powell or his room. It wasn't long before they homed in on Donald Harvey. Other hospital employees volunteered to take polygraph exams, so Harvey volunteered as well, after buying a book on how to beat the lie detector.

On the day of his scheduled polygraph, he called in sick, so was subsequently brought in for questioning. During interrogation by Detectives James Lawson and Ronald Camden, Harvey finally admitted to having put cyanide in Powell's gastric tube because he felt sorry for him and didn't want to see him suffer.

The detectives secured a search warrant for Harvey's apartment, where they found jars of cyanide and arsenic, books on poisons and occultism, and a detailed diary account of the Powell murder. On April 6, 1987, Harvey was indicted for aggravated murder in the first degree in the death of John Powell. He pled not guilty by reason of insanity and was held under a $200,000 bond. William "Bill" Whalen, a former assistant Hamilton County district attorney who had gone into private practice, was assigned by the court to defend him.

A competency hearing the next month heard the testimony of a psychiatrist and psychologist who had examined Harvey. Both concluded that though the defendant had a history of depression, probably derived from childhood experiences, he knew right from wrong, was not psychotic, and was free of any defect of mind. Just as important, Harvey told Whalen that despite his initial plea, he did not want to go with an insanity defense. Whalen then determined he would aim for a light sentence based on the fact that it was a mercy killing, Powell had been unlikely to emerge from his coma, and rightly or wrongly, Harvey perceived that he was doing the family a favor.

But Bill Whalen's task became more complicated when he received a call from Pat Minarcin, a news anchor at WCPO-TV, then the CBS affiliate in Cincinnati. After Minarcin speculated on the air with another reporter about whether it was possible that Harvey was responsible for other deaths at Drake, he began receiving calls—many of them anonymous—from nurses and other hospital workers about questionable deaths they thought Harvey might have caused.

Minarcin met with these sources and undertook his own investigation, correlating the number of deaths at the hospital when Harvey was and was not working. But he didn't want to make it public for fear of compromising the positions of the employees who had given him information. Weighing his options, he decided to contact Whalen and tell him that WCPO was considering running a story outlining a pattern of hospital deaths that could be connected to Harvey.

"I went straight over to the jail to talk to Donald," Whalen later told the *Enquirer*. "I asked him straight out, 'Donald, did you kill anybody else?' "

Harvey admitted that he had. Whalen asked how many.

Harvey said he couldn't tell him.

Whalen became irate. "You've got to be honest with me!" he demanded.

"You don't understand," Harvey said, explaining that he was not avoiding the question. "I can only give you an estimate." He figured it might be around seventy.

"When I heard him say the word *estimate*, I knew I was in trouble," Whalen recalled in a newspaper interview. The attorney suddenly found himself in an ethical dilemma. He had a solemn responsibility to defend his client to the best of his widely acknowledged legal ability. At the same time, he was utterly revolted by Harvey's wanton crimes and knew he should never be free to hurt anyone else. Ultimately, he decided that his primary task was to keep Donald Harvey out of the Ohio electric chair.

Whalen contacted Joseph Deters, from his old office with the Hamilton county prosecutor. Together, Whalen and Deters sat across a table from each other as they listened to Donald Harvey describe what he had done. He told Deters that there had been other cases and offered the government a deal: Harvey would plead guilty if County Prosecutor Arthur M. Ney Jr. would take the death penalty off the table. The Ohio law was that two murders were required to make the defendant eligible for execution, a threshold Harvey had unofficially crossed many years ago.

Whalen knew that the state would have great difficulty substantiating all of the deaths through investigation because there were no eyewitnesses and, in most cases, no autopsies or other postmortem examination had been performed. That would leave a lot of relatives, friends, and survivors in limbo, not knowing whether their loved ones had died of natural causes or murder. Both attorneys knew that this latest development could also potentially pave the way for a multitude of wrongful death civil suits.

Deters was convinced the investigators could find enough evidence to prove another murder without Harvey's confession and identified medical and scientific experts in surrounding states to examine the cases in which Harvey was suspected.

To keep the pressure on him for a deal, Whalen gave Ney a deadline and kept up his alliance with Pat Minarcin to maintain his position.

Whalen and Ney worked out that Harvey would plead guilty to twenty-eight counts of murder and offer a full confession and would receive three consecutive life sentences. Under the parole laws at the time, it meant that Harvey would not be eligible for release until he was in his mid-nineties.

The confession, Deters told us, lasted twelve hours and was extremely detailed, as Harvey methodically plodded through each of his kills. It was so overwhelming that those listening to it doubted he could have done everything he said, and perhaps was just bragging. But to check, the authorities exhumed twelve bodies that would have had recoverable poisons, according to Harvey's account. As Deters commented, "Every fact he told us checked out."

Ultimately, Whalen came to believe that Harvey had killed sixty-eight people.

With the Ohio deal in hand, Whalen went to Laurel County, Kentucky, commonwealth attorney Thomas Handy and got a similar arrangement for confessions of nine murders at Marymount Hospital. Circuit Judge Lewis Hopper handed down life sentences concurrent with those in Ohio.

Even after sentencing in Ohio and concurrent prison terms in Kentucky, Harvey kept "remembering" other murders. As a result of his plea bargain with Hamilton County, if he didn't fully cooperate with the ongoing investigation, charges could be brought in the newly uncovered cases that could bring the death penalty.

"I've been portrayed as a cold-blooded murderer, but I don't see myself that way," Harvey told the *Lexington Herald-Leader*. "I think I'm a very warm and loving person."

18

THE MAKING OF A KILLER

One of the reasons I was quite interested in interviewing Donald Harvey was that I had actually been brought in to consult on the case back when he was arrested. Since the VA hospitals are federal installations, the homicides committed there gave the FBI jurisdiction, and they had investigated Harvey's places of employment and all his personnel records.

At Quantico, I received a call from the special agent in charge—SAC, in bureau parlance—of the Cincinnati field office requesting that I provide on-site consultation in interview/interrogation strategies that would be most effective when they interviewed Donald Harvey. I flew out to Cincinnati that same day. When I arrived at the field office, I met the SAC as well as the two agents assigned to the case, who provided background on Harvey and the case details, similar to what we've just laid out. In many previous instances when it came to who would or should conduct the interrogation, I would be asked to select the agent I thought best suited for the task, based on personality or demeanor. At times I suggested the SAC conduct the interview himself (always a him in those days) because the suspect would respect that authority. I was sometimes asked to conduct the interview, but I preferred the coaching role, plus, if I were to get involved in the interview process, I would be spending most of my days in court testifying and that was not my mission.

When I'd completed my review, I met again with the agents and suggested a nonconfrontational, soft interview approach. Regardless of their personal feelings, they should make it appear they were sympathetic, that they knew each of his victims was going to die anyway and he was merely putting them out of their misery. This approach, I hoped, would keep him talking and, in the process, reveal his true intent with each killing. I instructed the agents to come across as empathetic and to provide him a face-saving scenario, suggesting that the crimes should be referred to as "mercy" deaths and for them to refrain from using such words as *murder* and *homicide*.

The agents assigned to this case were about the same age as Harvey. I suggested interviewing Harvey at the FBI field office, which would indicate power: power the FBI had, and Harvey no longer did.

The room where Harvey would be interviewed had a two-way mirror so that I and others could watch the interview and make suggestions to the interviewing agents.

Harvey did not have a criminal record at the time and the FBI was still in the process of investigating him, so we didn't have very much to go on. What we learned, though, was that Harvey seemed relieved that he was identified and arrested and appeared rather nonchalant about the whole thing. Seeing that, I told the agents there was a good chance he would speak freely about his involvement. What was not known at that time was the number of patients he had killed, so the face-saving scenario I suggested was still in play.

Harvey was brought into the interrogation room and appeared friendly, with a "you got me" sly look on his face. The agents asked him in-depth questions about his personal background before leading up to the crimes.

It soon became clear to me that we would not have to use any sophisticated interview techniques because Harvey wanted to talk. It was almost as if he were onstage delivering a soliloquy in a play.

This would be the first of multiple interviews the agents would conduct with Harvey. I would have liked to interviewed Harvey for my own purposes from a behavioral perspective, but the case was so complex there was no time for that. I thought perhaps someday I might have that opportunity.

DONALD HARVEY, INMATE NO. A-199449, ACCLIMATED QUICKLY TO LIFE AT THE Southern Ohio Correctional Facility in Lucasville. He was a cooperative prisoner with a clean disciplinary record and even agreed to participate in a training video created by a health-care security company on how to prevent others like him from working in hospitals.

The Lucasville facility does not have the Gothic oppressiveness of many of the prisons I have visited. Located near the Kentucky border, it looks more like a modern brick factory complex, possibly a pharmaceutical manufacturer, with a central guard tower that in a different setting could be mistaken for an air traffic control tower. But that relatively benign appearance should not fool you. The prison is home to some of the most violent offenders in the state, as well as the Ohio death house.

When we convened for the interview, I told him that though we had never met, I had been consulted when the FBI was preparing for their meeting and interrogation. I had been in an adjoining room watching the interrogation to see if I could help the interviewing agents get any information out of him, but as it turned out, I conceded, Harvey was so forthcoming on his own.

Based on this and other previous behavioral patterns, I expected him to be about 90 to 95 percent truthful with me. It was the other 5 to 10 percent that interested me.

The room was about eighteen by thirty feet and well lit. Harvey greeted me with a big smile as we shook hands. I returned his smile and did not apply much pressure when I grasped his hand.

"So you were behind the wall?" Harvey asked me in his soft-spoken southern accent, an amused smile on his face. He was still an innocuously pleasant-looking man with a crew cut, a mustache, and plastic rim glasses—softer, rounder, and older than the handsome, curly-haired guy he'd been at the time of his arrest.

Knowing that he had taken part in the training video, I emphasized what a great service he would be performing if he were equally as forthcoming about his background—his childhood, school, and formative experiences—and his advice to help others from going astray.

Throughout our preliminaries, he continued to appear charming and accommodating, as if he wanted to ingratiate himself to the entire film crew there with me. The first thing he asked was to meet the producer, Trisha Sorrells Doyle, who had made all the arrangements for the interview and filming. Trisha is an Emmy Award winner who has worked for several major television shows, including *60 Minutes*, *ABC News Primetime*, and *20/20*, and is a Columbia Law School graduate. She is also very attractive.

"Oh, I am right here. Hi," she said.

"Hi," Harvey replied. "You are the best stalker I have ever met. I am giving the interview because you have stalked me for four months."

I agreed that she was good, and Trisha went along, thanking him for the compliment and jokingly asking if he would pass along the comment to her boss. Harvey then told me I should recruit her to the FBI. Even though he was gay, I got the impression he was flirting with a pretty woman to show us what a normal, regular guy he was.

Prison officials had already told me that Harvey had wanted to make sure his clothes were cleaned and pressed so he could look as good as possible. In fact, looking at old photos from a court appearance, he was so meticulous in his appearance and grooming that it is difficult to tell the lawyers from the defendant. After our interview, he wrote a letter to Trisha on flowered stationery.

What intrigued me about Donald Harvey was that most serial killers are constantly on the hunt, some of them almost nightly. The hunt is a key component of the fantasy and often just as fulfilling as the crime itself. I have likened the serial predator to a lion on the Serengeti Plain in Africa, looking out over a large herd of antelope at a watering hole. Somehow the lion locks on a single one out of those hundreds or thousands of animals. He's trained himself to sense weakness, vulnerability, anything that makes that one individual a more likely or preferable victim than the others. It is part of the hunter's essential psychic makeup.

But Harvey was more like a lion in a zoo. He didn't have to hunt for his prey; it was "served up" to him daily. So what was his thrill? What was his satisfaction? What substituted for the hunt itself? Or in his mind, was there still a hunt within the confines of the hospital?

Applying the nature-versus-nurture question here, one has to ask whether Harvey fits into the "made" or "born" category. As we've noted, there seems to be some organic predisposition, and we don't know enough about neuroscience yet to understand it. Is it genetic? Possibly. But when we study the background of a violent predator, we generally find a sibling who has no such tendencies. Or we find one who has some of the same tendencies but manifests them in a diametrically opposite manner.

A fascinating example of this is Theodore Kaczynski—the Unabomber—and his younger brother David. Both young men were extremely bright, both lived alone for a time in primitive surroundings, both had visions of improving society. Yet Ted became a lethally effective serial bomber and David became a social worker and a Buddhist whose deeply felt ethics motivated him to turn in his brother once he and his wife realized that the published "Unabomber Manifesto" sounded disturbingly like Ted. The only deal David wanted for his cooperation was that the death penalty be taken off the table for his brother, which speaks volumes about how different these two brothers who emerged from the same background turned out. Ted was ready to murder the innocent. David couldn't countenance executing the guilty.

Gary Gilmore, the multiple murderer in Utah referred to earlier, who was the first person executed after the U.S. Supreme Court lifted its ban on capital punishment in 1976 and is the subject of Norman Mailer's Pulitzer Prize–winning epic, *The Executioner's Song*, has a brother named Mikal who is a distinguished journalist, writer, and music critic.

Our research suggests that the *combination* of hardwired neurological factors and a bad childhood and adolescence contributes most often to an antisocial personality. It is possible that without one or the other influence, the violence-prone predator never emerges, as suggested by our informal control group of law-abiding siblings like David and Mikal. But this is not a laboratory experiment where we can play it out two ways. At this point in the development of both neuropsychology and criminology, the best we can offer is theories.

Let us also note that everyone calibrates a bad background differently. For Joseph McGowan, the controlling, dominating mother seems to have been enough to put him over the edge when she rejected his intention to marry. Joseph Kondro felt dislocated by knowledge that he was adopted, but more likely, lack of good judgment and impulse control played a larger role in his development.

Donald Harvey actually shared a childhood trauma with his mother, who was sexually assaulted when she was twelve and developed partial paralysis, probably functional rather than organic, as a result. She married Donald's father when she was still a teenager and he was twenty-nine. Harvey believes she looked up to him as a father figure, which does not sound farfetched.

With Donald, whatever his organic predisposition, the formative years seem to have played a crucial role in his psychic development. Once we got beyond the small talk, the first topic I brought up was the abuse he had suffered at the hands of others, beginning, as I understood it, at age four.

In his gentle, matter-of-fact manner, Harvey confirmed that this was the first sexual abuse he recalled—at the hands of his uncle, who was only nine years older. When Donald was about five and a half, he told me, a "neighbor man started messing with me, too." His uncle told him that this was a thing that boys did, and the neighbor said that if Donald told anyone, he would shoot his mother and father and the young boy would have to go to an orphanage. Harvey also claimed that his uncle was later into "wild sex" and beat both his first and second wives. Harvey didn't remember the first wife too well because he was very young at the time, but he had indulged in threesomes with his uncle and the second wife, and this was how his uncle "could get it up."

Now, I never condone or excuse violent or harmful behavior, and an abuse-filled childhood is not an acceptable rationalization for a violence-prone adulthood. But if we consider these two unwanted relationships early in Harvey's life, not to mention seeing a man beat his wives, it is not difficult to see how he might have grown up with some significant negative feelings toward those he perceived—rightly—to be more powerful than he and the authority figures who allowed these traumas to be visited upon him.

Of course, I don't know whether these two child molesters contributed to young Donald's sexual orientation, but I tend to doubt it. The science is pretty convincing that male homosexuality is hardwired in an individual, probably before birth. But what is interesting is how Donald "got back" at these two serial child rapists. Rather than turn them in or exact revenge on them, as he matured into his late teens, he told me, he decided to become an equal in both relationships. He was going to demand his own dignity. In fact, he told me that he lived with his uncle and his second wife for a year when he was seventeen and eighteen, before he went to work at Marymount Hospital.

Harvey did not display much of the childhood behavior we often see in serial killers, which didn't surprise me much because, despite his horrifying death toll, he wasn't like most serial killers. Notwithstanding his youthful traumas, he was not a bed wetter. He didn't start fires. And except for two incidents, he did not harm or show cruelty to small animals. In one, his mother told him to return a baby chick to a neighboring farm and Donald wrung its neck instead.

When I interviewed Richard Speck, who had raped and murdered eight student nurses in their shared apartment in Chicago in 1966, he told me that he had cultivated a sparrow as a pet in his prison cell. When authorities told him that he couldn't keep the bird, in front of them he proceeded to squeeze the bird to death in his hand and then threw it into an open fan. It was all about power and control. I saw a similar emotional scenario here with Harvey.

After the chick incident, when Donald was about sixteen, he led two of the neighbor's cows into the woods and slit their throats. He said he did it not because he wanted to hurt the cows, but because he wanted his neighbor to suffer a financial loss. Recall also that Donald grew up in an agricultural community in which slaughtering chickens and cows by the methods he chose was a normal part of life. The reason he slaughtered these three animals was what was abnormal.

I questioned Harvey on his opinion of nature versus nurture. His answer was opposite from Kondro and aimed right back at me.

"So, you are basically [asking]: Is someone born a bad seed? Well no—just think of circumstances. They could have a good . . . I mean, who says *you're* not a serial killer? I mean, you have a fascination with going to talk to them and stuff like that. You might be like Hannibal Lecter, we don't know."

It is true that author Thomas Harris visited my FBI unit and spoke to my colleagues and me extensively before he wrote *Red Dragon* and *The Silence of the Lambs*. But even if for the sake of argument we accept Harvey's premise, a major difference between us is that despite my "fascination," I haven't gone out and murdered anyone. And the high percentage among serial killers of wannabe cops or men who failed to get into the police force speaks to nothing more than their lust for personal power and dominance, rather than a higher calling to serve the public and keep the peace.

"What you're saying," I replied, "has been the response that I give;

that I don't believe people are bad seeds who go out and kill. I think extraordinary things happen, but then it's up to that individual—how you are going to respond to the stresses in your life. You can withdraw, you can cope with it, you can succeed, or you can externalize and aggress against others."

Curiously, Harvey rejoined, "My family was religious on both sides. And my mom and dad went to church, basically every Sunday."

"You did, too," I said, recalling that Joe Kondro had become "real spiritual" in prison and Dennis Rader was the immediate past president of Christ Lutheran Church of Park City, Kansas, at the time of his arrest.

Again, Harvey showed some pretty pointed insight. He acknowledged that he attended church regularly as a young person but said that it was because the neighbor women served good food afterward, better than he got at home. He said he would do small chores for these women and they would give him candy and biscuits and gravy, which he loved. "I learned how to manipulate at a very young age," he said, and that talent affected every aspect of his life, including school, where he made a point of becoming the teacher's pet.

He freely admitted to the murders, but when I asked him about the first one—Logan Evans—he had his psychological excuse already prepared: "According to the psychiatrists and stuff I have talked with, I wasn't going to take anymore. I mean, I had had enough, and when he put the stool in my face, that was the breaking point."

"Did you think that that's what happened, because the psychiatrists told you?"

"It started me on the roller coaster which I couldn't get off."

This is probably the most common theme that has emerged in all of my confrontations with killers across the table. There is almost always an external reason the killing started. As a result, nearly all serial killers believe their crimes are justified, or at least explainable. They perceive themselves as the true victims—yet another manifestation of their extreme narcissism. And if a prison psychiatrist happens to hand you the excuse—well, so much the better.

And the manipulation Harvey had learned so well never stopped. He told me how two gay orderlies introduced him to the funeral director in London, Kentucky, who was married with three children and bisexual. According to Harvey, the man liked to have sex in coffins and ice-filled bathtubs. He was also the one who got Harvey into matters of the occult.

I asked Harvey if he participated in the coffin and bathtub sex.

"I did the ice; that was cold for me. I mean, I am sorry. And I had cold showers and stuff like that and he had all kinds of . . . He was into, I guess you call it devil worshiping, of a certain type. He liked to shower blood. He stole body parts from the corpses and he would always get the body parts from somebody killed in a car wreck or something like that."

That all sounds pretty creepy, even for a guy like Harvey. So why did he go along with this weirdo?

"He showed me how to use the plastic bag. He showed me how to use different things and how, you know, if they did an autopsy, forensics could find fiber or whatever. And what marks the pillow would leave—a certain mark as opposed to the bag, when you held the bag down or put it in the mouth and nose of a comatose patient."

This high school dropout was still learning all the time.

When Harvey brought up devil worship, I brought up "Duncan," whom I'd read about in his file.

Many of us had imaginary friends as young children. But I have come across a fair number of serial predators who had something similar as adults, or at least claimed they did. These "friends" took several forms. The ones with which prosecutors are most familiar are those supposedly caused by multiple personality disorder.

In real life, MPD—now often referred to as dissociative identity disorder—is an extremely rare phenomenon, and generally shows up in young children who have been severely abused and "escape" into other personalities that are stronger and/or detached from the actual personality. These cases are very real and heartbreaking.

Perhaps the most famous case of MPD in an adult was Christine

"Chris" Sizemore, whose life was depicted in the 1950s book and movie *The Three Faces of Eve*. In the 1980s, Sizemore came to speak to our Criminal Psychology classes at Quantico to help us understand what actual multiple personality disorder was all about. She told us that at some points she had as many as twenty different personalities and felt they were with her since birth. She made a sensitive and convincing case.

But much more often, when those of us in law enforcement see a claim of MPD, it is post-arrest. Though the suspect/defendant may never have given any indication to those around him that he has more than one personality, if the evidence against him is strong and there is no other way to explain his action, he or his attorney will put forth a multiple personality disorder defense. In other words, while his "body" may have committed the murder, it was another personality working within that body that had the motive and mens rea (literally, "guilty mind"). Legally, both the mens rea and the act are necessary components to make up a crime.

Larry Gene Bell abducted and killed seventeen-year-old Shari Faye Smith in Columbia, South Carolina, in 1985, after first allowing her to send a heartrending "Last Will and Testament" to her agonized parents and sister. When he was identified and captured in a fulfilling collaboration of FBI profiling and fine police work, I took a crack at interrogating him to see if we could get him to confess. I played the MPD card, and before long, he conceded that while the "good" Larry Gene Bell could never have done such a thing, the "bad" Larry Gene Bell might have. I didn't much care after that what happened to the good Larry Gene Bell, but I can't say I wasn't gratified when the bad Larry Gene Bell was electrocuted on October 4, 1996.

The sadistic killer Kenneth Bianchi, who with his cousin Angelo Buono Jr. murdered at least ten young women in the 1977–1978 Hillside Strangler cases in California, tried to plead insanity due to a second, malignant personality during his 1983 trial. He convinced several court-appointed psychiatrists he was for real. My colleagues at Quantico and I had consulted with the distinguished psychiatrist Dr. Martin Orne of the University of Pennsylvania, who was doing pioneering work on hypnosis and memory distortion. At Bianchi's trial, Dr. Orne testified that true MPD sufferers tended to have three or more distinct personalities. The next day Bianchi suddenly developed another one named Billy that he had never mentioned before.

Which, for the moment, brings us back to Donald Harvey. At certain times, he claimed to be motivated by a satanic spirit named Duncan, whom he would consult before some of his killings. He would set up candles representing possible victims, and if a candle flickered, it was Duncan's indicator that this individual should be killed.

First, as I've indicated, my FBI colleague Ken Lanning effectively demonstrated that satanic ritual murder as a motivator does not really exist, so if the Duncan story had been true, this would be so unusual as to be truly noteworthy in the annals of crime. But aside from that, it just didn't add up from a forensic standpoint. Many, if not most, of these murders were crimes of opportunity, and some of them were out-and-out spontaneous, so it didn't make any sense that Harvey would be providing an array of potential victims for Duncan to pass judgment on. My instinct was that Duncan was just as bogus as Kenneth Bianchi's Billy and the 3,000-year-old demon dwelling in neighbor Sam Carr's black Lab that David Berkowitz claimed forced his murders of young couples in New York City.

When I had interviewed Berkowitz in Attica, he had tried to pass the dog story off on me. "Hey, David, knock off the bullshit," I said. "The dog had nothing to do with it." In fact, the story didn't even come into being until after his arrest. He laughed and nodded and conceded that I was right.

I have seen this trait of setting up an avatar to take the rap with some frequency in my career.

In his book, originally titled *If I Did It*, subsequently published as *I Did It* when the Goldman family won the rights based on their civil suit, O. J. Simpson recounts the killings of his ex-wife Nicole Brown and her friend Ronald Goldman as if O.J. had actually committed the crimes. From my perspective of forty-plus years in law enforcement and behavioral analysis, this book, written years after O.J.'s acquittal for the murders, was just another display of Mr. Simpson's contempt for moral standards, his sense of power over and remaining anger at Nicole. In other words: the actions of a sociopathic narcissist.

But he also brings an avatar into his retelling. Describing what happened on the night of June 12, 1994, Simpson writes that his friend Charlie—a name that prior to this book has come up nowhere in the investigation—tells him, "You wouldn't believe what's going on at Nicole's house," meaning her consorting with men. Freudian psychiatrists would call Charlie O.J.'s id, the emotional, impulsive part of the personality. My colleagues in law enforcement and I would simply call it criminal intent. This use of an imaginary entity not only distances the offender from the violent act but allows him to depersonalize the victim and what he has done to her.

Harvey didn't have Charlie, but he did have Duncan.

"So, Duncan, this out-of-the-world guy . . ." I began with Harvey.

"You better watch Duncan. He might get you," he replied with a mischievous grin.

We clarified that Duncan had come along in Donald's early twenties, so he couldn't be responsible for the early murders. "So, you can't blame Duncan?"

"Well, Duncan has been, probably was with me all my life."

"Do you think so?"

He smiled again and appeared to be toying with me. The idea of being able to manipulate a former federal agent was clearly very appealing. "Didn't you have an imaginary friend?" he asked. "Then how are you going to play cowboys and Indians? You want to be a cop? Now, you were a federal man, right?" He seemed to be implying that we grow up to become whatever we imagine as children. This is not an outlandish proposition, though as a child I wanted to be a veterinarian and never considered a career as an FBI agent. Was Harvey trying to suggest his imaginary friend Duncan had come along before his early adulthood and had motivated his choice of "careers"?

I brought him back to the night in London, Kentucky, when he became seriously inebriated and confessed to his first fifteen killings. I asked him if it was true that no one believed him.

"No one believed me," he confirmed. "They said I had a very active imagination. They sent me to see a psychiatrist, asked me if I was suicidal, was I drunk, was I smoking marijuana. No; first time I ever drunk liquor and I spilled my guts. So I am real trustworthy."

The irony of this to someone like me is that I am always skeptical of what convicts tell psychiatrists or other therapists, since they have something definitive to gain by convincing the therapist that they are cured, harmless, reformed, see the light, etc. And I am critical of therapists in criminal justice environments who believe these guys too easily. Here we had an opposite situation. Here we had a guy who *should* have been a convict, and law enforcement personnel and therapists who *should* have believed he was telling the truth, and instead they all concluded that he had "a very active imagination." And by not believing him, they set him free to keep on killing.

What is the lesson here, if any?

Normal people find it difficult to grasp the reality that predators really do think differently. We tend to want to evaluate them from the point of view of our own experience and life values, and then try to figure out what it is that "went wrong." In other words, what is the aberrant piece that once identified and "fixed" will make them think "normally" again? Well, in many cases there is an aberrant piece that either determines or influences behavior. But by the time some individual acts on his predatory urges, it is usually so completely assimilated into his entire personality that you can't simply take it out and replace it as you can a defective mechanical part. That is why the concept of rehabilitation is so problematic for violent offenders.

Once the damage is done, it is often all but impossible to repair it.

19

"I HAVEN'T CHANGED A BIT"

As disturbing as Donald Harvey's record is, there is perhaps another element to his case that is of overwhelming concern—namely, that he was able to kill undetected in a supposedly safe institutional setting for so long. How was he so successful at this for so long? Aren't there any safeguards and security measures—even some form of pattern recognition—built into the system? The answer is that he understood the system; in fact he understood it better than the people who administered it.

He explained that by watching the people around him and noting their habits and routines—who was attentive and who was lazy and didn't follow up—he could feel comfortable with what he was planning and doing. "You got people like me that sit back and just watch everything. Okay, those people like me—they don't have a care in the world. But those are the kinds you sometimes have to watch."

On any given ward, he could even say which side of the hallway was attended to by an RN—a registered nurse—and which side was attended to by a licensed practical nurse—LPN—and knew that he was likely to be asked to help out on the LPN side.

In other words, he turned himself into a profiler.

"When you go into a hospital to work, you look at the people. Go to the favorite watering holes and stuff; see what kinds of uniforms they use, how they act, how they relax. They go in, they throw their stuff down. You can snatch it and go, look and see what kinds of uniforms are on that day; do they have multi-uniforms—those are great! That means you can get your multi-uniform and just wear it. They may assign you five or six different colors. Right today, I can go right through any hospital I wanted. Just give me a day." I think what he was referring to was that in many hospitals, each position has a different color and/or style of uniform, so he could morph from one position to another simply by knowing which uniform to grab.

"I am sure you can," I observed.

" 'And how are you today?' You know—da-da-da, where is this? Or, 'Oh, I haven't seen so-and-so today.' And they say, 'Oh, yeah, so you are . . . ?' They don't know who you are talking about—Miss Johns, Miss Smith, whatever."

"Because you have familiarity?"

"You got to be able to talk a junkyard dog out of a garbage can."

"You are in a comfort zone, working in hospitals." Just about every predator works within his comfort zone. A comfort zone can be a geographical area—close to home or somewhere familiar to him. It can be an interpersonal zone, such as Joseph Kondro's comfort with the daughters of his friends. Or it can be a particular environment, such as Harvey's hospitals, where he knew he would find an unending supply of helpless victims.

THE PRESENCE OF KILLERS LIKE HARVEY AND THE VULNERABILITIES IN THE INSTITUtions of care are laid bare by the fact that there are plenty more killers where Harvey came from. Both before and after Harvey's reign there have been examples of killers within the health-care system targeting those who depend on them and give them their trust. The individual who probably holds the record for the most medical murders was not a nurse or orderly, but a physician—Dr. Harold "Fred" Shipman of Manchester, England. He is noteworthy in another way as well. Shipman goes beyond our previous cases in terms of motive, plus the fact that most of his crimes took place outside a hospital setting.

Harold Frederick Shipman was born into a working-class family in Nottingham on January 14, 1946, the favored middle child of a domineering mother who instilled in him a sense of superiority. When he was in his teens, his mother contracted lung cancer, and he became interested in medicine while taking part in her care. He was devastated when she died and he enrolled in Leeds University two years later to pursue a medical education. He married his wife, Primrose, when he was nineteen and she was seventeen and five months pregnant with their first child.

After medical school and internship, he joined the Abraham Ormerod Medical Centre in Todmorden, West Yorkshire, as a family practitioner, where he did well until he became addicted to the painkiller Pethidine, for which he was caught forging prescriptions. He entered a drug rehab program, paid a fine for forgery, and joined the staff at Donneybrook Medical Centre in Hyde. He stayed for nearly two decades, during which time he became known as a hardworking and dedicated physician, popular and trusted by his patients. In 1993 he opened his own office at 21 Market Street in Hyde.

A local undertaker noticed what seemed like an unusual number of deaths among Dr. Shipman's patients, all of whom seemed to die fully clothed and either sitting up or reclining. He questioned the doctor, who assured him there was nothing to worry about. One of his colleagues, Dr. Susan Booth, noted a similar pattern and alerted the local coroner's office, which in turn notified the local police. A secret investigation followed, but nothing stood out.

He seemed to concentrate on older or elderly women as patients, and made a habit of house calls, which his patients appreciated.

With each of these medical murderers, there is always one case that brings them down, either because of sloppiness or haste, special circumstances such as a legal requirement for an autopsy, or one individual who sees something that doesn't fit and won't give up the hunt.

Kathleen Grundy, former mayor of Hyde and widow of a university professor, was a sprightly and active woman of eighty-one who was found dead in her home not long after a visit from Dr. Shipman. Her daughter, Angela Woodruff, stunned by her mother's sudden death, asked Shipman if an autopsy should be performed before taking the body to the funeral home, but the doctor told her it wasn't necessary. He listed her cause of death as "old age" before signing the certificate.

Angela was an attorney and handled her mother's financial affairs. She was shocked to find that Kathleen had left a will dated more recently than the one she knew about, and that it left the bulk of her estate to Dr. Shipman. That was when Angela started suspecting him as a murderer. She contacted the local police and shared her ideas with Detective Superintendent Bernard Postles, who came to a similar conclusion. He secured a court order to have Mrs. Grundy's body exhumed. A postmortem examination revealed a large presence of diamorphine, a powerful form of the drug used to control pain in terminal cancer patients, administered within a period that coincided with Dr. Shipman's visit.

A search warrant on Shipman's house revealed medical records for many patients, a collection of odd pieces of jewelry, and the typewriter on which Mrs. Grundy's surprise will had been written.

Police reviewed the death records of all of Shipman's elderly patients, and then focused on those who had died at home and those who had not been cremated, so that a body could still be examined. Not surprisingly, Shipman had encouraged many of the grieving families of his patients to opt for cremation, thereby destroying the evidence of his crimes.

A careful examination of patient records, correlated with computerized time stamps, revealed that Shipman had altered the charts of many of his victims so that their conditions matched more closely with the apparent causes of death, which accounted for the lack of conclusive evidence in the initial investigation. As a result of this one, though, plus exhumations and autopsies on a number of possible victims, Shipman was charged with fifteen counts of murder.

The trial began on October 5, 1999, in Preston Crown Court in Lancashire. Not surprisingly, the defense tried to assert that any deaths Dr. Shipman may have caused were due to his compassion for their extreme suffering. Sound familiar?

The prosecution countered that Shipman was in love with the power of life and death, and that none of the alleged victims was suffering from a terminal illness.

After presenting its medical evidence, showing that in most cases the victims had died of morphine toxicity, the prosecution proved through fingerprint and handwriting analysis that Kathleen Grundy had never even touched the will she was supposed to have executed.

On January 31, 2000, after six days of deliberation, the jury found Dr. Harold Shipman guilty of all fifteen counts of murder and one of forgery. The sentence of what in Britain is referred to as "whole life tariff" effectively precluded any possibility of parole.

After the trial and sentencing, a two-year clinical audit chaired by high court judge Dame Janet Smith and known as *The Shipman Inquiry* concluded that over a twenty-four-year period, Shipman might have been responsible for at least 236 patient homicides. This made him easily the most prolific serial killer in United Kingdom history. About 80 percent of his victims were women. Shipman arrogantly continued to maintain his complete innocence, while Judge Smith stated that her commission had investigated many more deaths for which there was no conclusive proof.

During the early morning check at the high-security HM Prison Wakefield in West Yorkshire, Shipman was found dead, hanging in his cell, one day short of his fifty-eighth birthday. He had used bedsheets tied to the window bars.

Killers like Shipman and Harvey have been able to take advantage of the systems in place, but opportunity alone isn't enough to get away with the crimes. But he—or in this type of crime, sometimes she—must also know how to behave within those systems.

"Always be nice—that's the biggest key," Harvey told me. "Be friendly, but don't be friendly to where it is suspicious friendly. Always be nice because the people treat you nice, too, a majority of the time." This is akin to a Ted Bundy–type sexual predator being outwardly charming despite his murderous intentions.

As we've noted with other serial predators, you can lock up the body, but the mind remains free to relive the crimes over and over and get back into the moments of the perpetrator's greatest criminal satisfactions. I asked Harvey what kinds of things he thought about when he was alone at night. Were there any specific murders that were particularly close to his heart, that he evoked in the midnight of his soul? His answer was surprising.

"I am building a house—a log house right now. I get this one magazine, and I build two, three—I mean, four bedrooms right now, but I only got three baths."

"You are doing it in your mind?"

"Yeah, yeah. Sometimes I write down stuff. Yeah, we [he and Duncan?] build, and then maybe next year it will be different—a church or something like that."

From most incarcerated felons, I would consider this response to be crap. But from someone who took killing as casually as Donald Harvey, it actually made sense. Unlike Dennis Rader, for instance, who spent much of his waking time fantasizing about binding, torturing, and killing entire families and even storyboarded many of his ideas with drawings, Harvey went about his business. And if his business included killing someone, it certainly made him feel better, but he didn't seem to dwell on the acts the same way. As he had just told me, he had made a sufficiently extensive study of his surroundings and coworkers that he could feel comfortable with what he was doing, so when the opportunity to kill arose, he could take it.

However, I wasn't willing to accept that he never thought about it or fantasized. I tried another approach. If building houses in his mind was his current fantasy, had killing helpless people been a dominating fantasy growing up or once he began working in hospitals?

He insisted the crimes were much more "practical" than that. "No, I mean, not like that at all. It is like Carl and all the people I did meet—I didn't let it go. I am like this: 'If you don't want to be with me, bye-bye; let the back door hit you.' "

Whether that "bye-bye" meant "Nice knowing you," or "Say goodbye to life," he left purposely ambiguous. But his criminal career had made clear that anyone who bothered him could easily become a mortality statistic.

He stated that he took the same attitude in prison—being nice to everyone, saying "Good morning," and going about his job of transporting dirty laundry and cleaning offices. I pointed out that he was in protective custody rather than the general population, but he said that was the authorities' choice, not his; he was fine getting along in the general population.

In his own way, Harvey was as practical-minded as Joseph Kondro, who had figured out that it was easier to gain the trust of a victim he already knew and felt he could get away with it if the police couldn't find a body. From working in the morgue and assisting with autopsies, Harvey not only knew the effects and residual evidence left by each drug or condition, but also that, depending on the apparent cause of death, the autopsy would be performed according to a particular protocol that might be unlikely even to focus on the actual cause of the murder.

"See, one way [of killing], you take a patient that's got heart disease—you take the spinal fluid and drain enough of the spinal fluid to cause him to have a heart attack. One lady, she was treated for pneumonia. I caused her to have the sputum plug, so basically she smothered to death."

If the patient had heart disease and seemed to die of a heart attack, there would be no reason for a pathologist to check the spinal fluid. And the woman suffering from pneumonia seemed to die of pneumonia, so there was nothing suspicious. On top of that, Harvey knew that unless there was a specific indication, most hospital deaths would not be subjected to autopsy examination.

I asked Harvey, "Going back retrospectively, were there any things you would have done differently with Powell, the patient that got you here?"

"Well," he responded, as if the answer were obvious, "I would have never gave him the cyanide; I would have cleaned out the G-tube better. I mean, you can't go back and second-guess yourself. I don't try to second-guess myself anymore."

And, of course, the nastiest deaths weren't really his fault. The one who died of peritonitis because Harvey had shoved a straightened coat hanger into his catheter—"that was poor security on the hospital's part," for letting Donald tend to him after he had hit Donald in the head with the metal urinal.

This reaction is not unusual. A rapist once told me when he saw a woman sitting on a bar stool in a restaurant in a short skirt and apparently no underwear that she was looking to get raped, and he merely obliged.

I pointed out to Harvey that what he was doing was known as projection. He allowed as how "what I did was wrong," but rationalized that he was eighteen years old and had not been told how to deal with situations like that. Ultimately, he was willing to split responsibility with the hospital: "No, they are not responsible for the death part; they are responsible for his security because they tied him down and told me to take care of him. I didn't tie him down; he was already tied down."

And remember how he agreed to take part in the hospital security film as a public service to prevent others from getting away with crimes like his? Well, when I asked him what he'd like to do if he was paroled before he was in his nineties, he replied, "Well, since you like to talk to serial killers and murderers, I think I will go work with you, because we make a good team. What they [the profilers] can't come up with, I can come up with the rest."

Try as I might, this was one guy I couldn't get to emotionally—to the point where he'd break down and show me his true vulnerabilities. His defense mechanism seemed to consist of just skating along the surface of everything.

After he was arrested and had been convicted of several murders, Ted Bundy used to get off on academic types asking him to help them understand his deep, dark criminal mind. Bundy was probably as close to the archetype of one species of serial killer as you could get: handsome, intelligent, charming, glib, and resourceful. And he brutally killed at least thirty young women from Washington State to Florida during the 1970s. The late crime writer and former police officer Ann Rule worked next to him in a rape crisis center in Seattle and never suspected his dark secret. In fact, she believed he actually helped people and saved lives working the hot line there.

Bill Hagmaier from my Investigative Support Unit went down to talk to Bundy in his last place of residence, the death row cellblock at the Florida State Prison at Starke. Talking to the FBI was probably the ultimate ego boost for him. Despite his God-given advantages, Bundy had had a troubled childhood and, like most serial predators who have been caught, was quick to place blame on externalities for his crimes—in his case, violent pornography. While the porn might have given him ideas and stoked his lust, had the murderous impulses not already been there, that would not have turned him into a killer.

Two things struck me in what we learned from Hagmaier's interview. First, the vicious sexual assault and murder of the beautiful young women who were Bundy's victims of choice were not the most important or satisfying elements of the crimes for him. What he described as really getting him off was the thrill of the hunt and capture, and then, like Dennis Rader, the sublime power of life and death over another human being. He related that when he abducted two women, Janice Ott and Denise Naslund, from Lake Sammamish in Washington State, he kept them alive as long as he felt he safely could and made one watch as he killed the other. This is what gave him the most sadistic pleasure. He told Hagmaier that the reason he killed so many beautiful young women was that he wanted to. He told Bill he enjoyed it as an almost mystical experience that allowed him to fully possess his victims. In that he was also like Rader, who was operating around the same time.

The other thing that impressed me about the Bundy interview was that he had ultimate confidence in his own ability to escape from any dilemma. This manifested as actual physical escape, as he achieved from every crime scene, as well as from the jail at the Pitkin County Courthouse in Aspen, Colorado, where he was being held temporarily after extradition from Utah to stand trial for murder. He was picked up eight days later, but escaped again six months later, shortly before the start of his murder trial. He was also stopped by a patrol officer while driving around Tallahassee, Florida, but as the officer walked back to his patrol car to run the license plate, Bundy bolted.

But that was not his only escape strategy. When he was finally picked up for real, he thought that by offering to "help clear" open murder cases without actually admitting guilt, he could bargain his way to a lighter sentence. He felt that by talking to the academics who wanted to interview him, he was making himself too "valuable" to be killed and therefore deserving of special treatment. When he finally did come up for trial, he arrogantly demanded to be allowed to act as his own counsel, merely "assisted" by Michael Minerva, a highly respected public defender.

When, toward the very end, he contemplated his ultimate fate from death row, he became desperate to do anything he could to stave off his execution. He begged both Bill Hagmaier and Dr. Robert Keppel, a criminal investigator from Washington State whose detective career had begun with the early "Ted murders," to intervene with the authorities and keep him alive. Finally, the forty-two-year-old Bundy's vaunted resourcefulness failed him. He was executed in the electric chair on January 24, 1989. Bill remained with him until the day before, showing him some of the human kindness Bundy had denied so many others.

Though not the same type of serial killer, Donald Harvey exhibited the same kind of warped sense of pride. The fact that people like me requested interviews clearly made him feel important.

I asked him how he'd feel if his mother passed away.

"She's not much older than me, so it's hard to say who will go first."

"I mean, it would be a really terrible thing."

"Sure, and it's terrible to lose any parent or relative."

Well, how did he feel when his father died?

"I was a young man."

And as far as his family?

"That's your family on the outside. This is my prison family."

Nor did he express any remorse about the murders: "I never looked at it as murder till I was actually arrested. I always looked at it as mercy, you know. Today they have the hospice and assisted suicide."

"They get the choice," I pointed out.

He agreed with that but countered that most of them were in no position to make the choice and didn't have any close family to make the choice for them. Interestingly, though, instead of characterizing what he did in terms of euthanasia or assisted suicide, he observed, "I was being the judge and the jury and the executioner." *Executioner* is not a word we normally associate with mercy killing.

"But they weren't all mercy killings," I said.

"No, they weren't all that."

"And just so we know the different methods again, you mentioned one time the pillowcase. What are the other methods?" I knew all his methods but wanted to see if he would differentiate between the victims he killed for "mercy" and those he killed for other personal causes.

"I used morphine," he began. "I used morphine because it wasn't controlled that much in Kentucky, especially in small hospitals in the

1970s. They had it in refrigerators. I used cyanide. I used arsenic. I used adhesive cleaner. Well, a plastic bag, too."

"Were you experimenting or was it just whatever was available?"

"I wanted the quickest method of death. And I did unplug a couple of ventilators. I used an adhesive cleaner because that will cause them to have a sputum plug and they would die of pneumonia. I used cyanide in IVs—I give it by injection. Cyanide on a black person will not show, but it will show on a white person. If you give it in . . . you don't give it by shot, you put that in the IV—no, I used it in their feeding tubes—that's what got me busted. I was in a hurry."

Note how dispassionate and procedural this all is to him.

I wanted to follow up on his sense of himself as a dispenser of justice. "What about this guy Nathaniel Watson?"

"He was a rapist."

"You *think* he was a rapist."

"No, he was. He was in the county jail."

I had looked into the case. "It was kind of questionable whether or not, in fact—"

"Well," Harvey interrupted, "he had been questioned in a rape of six women and he had a stroke. And every time he looked at a white woman, he had an erection. He looked at the black women or black men—no. Even when he looked at the doctor, if he was white, he had erections, so I killed him. I mean, he was in bad shape; really he was in bad shape."

There was no credible evidence that Watson was a rapist, but Harvey clung to the belief because it provided him a convenient excuse for the murder.

"So, you felt it was fine to take him out?" I asked.

Then he took an opposite tack, describing how Watson's comatose body was unnaturally contracted, that he had chronic bedsores, and that it required two people to move him. Harvey compared him to Karen Ann Quinlan, the young Pennsylvania woman who lost consciousness in 1975 after consuming several drugs mixed with alcohol and lay in an irreversible coma while her religious family tried to have her removed from a respirator and returned to her "natural state." Even after an appeals judge sided with the family and Ms. Quinlan was removed from the respirator, she lived for almost a decade with just feeding tube support. This was the case that brought right-to-die issues to national attention. Harvey claimed he didn't want Watson to suffer such a fate.

So, this was a twofer—Harvey killed him because he was a rapist and/or because he was a hopeless stroke victim. I bring this up to highlight the contrast between Harvey's highly logical, methodical analysis of his surroundings and colleagues and his totally muddled and self-justifying reasoning on why he kept killing helpless people.

What was becoming increasingly clear was Harvey's profoundly ambivalent and immature orientation toward other people. It was as if he were acting out on a preadolescent level, fantasizing about—and in his case acting against—anyone who annoyed him at a given moment, even if he generally had an okay or neutral relationship with that person.

This was driven home when I asked him about the neighbor he had tried to poison, as well as infect with HIV and hepatitis B.

"Loves gay guys," he responded. "And that's all she has—all of her friends were gay men. She felt safe with them. She was a good cook and she was actually nice most of the time."

So why did he try to kill her? Because he thought she was telling stories about Donald cheating on Carl with other women. Apparently, she resented that Carl was spending more time with Donald than with her. And despite the fact that "she was actually nice most of the time," Harvey's only real reaction when I brought her up was that arsenic would have been a better and more efficient way to kill her.

Carl, on the other hand, he did not want to kill. Harvey confirmed that he gave him only enough arsenic to keep "his penis in his pants and stay out of the parks and stuff."

After a while, though, that wasn't enough to keep Carl in check.

"Finally, it got so bad with him on a sexual thing—he'd still go out with diarrhea and be puking, and he'd still go out."

I went through several more of his non-hospital kills.

Neighbor Helen Metzger: "She was a very nice lady. Carl stole about $100,000 from her and she was going to the cops and I couldn't afford to let her go to the cops because I had been questioned and I didn't want to mess up stuff with me at the VA."

On the subject of mercy killings, he said that the worst thing was when he was put on the oncology ward, because so many of the patients were near death and had no family, so he essentially appointed himself to take them out of their misery. Had he stayed in general nursing, he thought, he would have been fine, which of course is totally inconsistent with the portrait of himself he had already revealed to me.

Harvey's simple, easygoing persona was coupled with a Bundy-like resourcefulness and criminal imagination.

When the Nathaniel Watson death aroused suspicion, he had an elaborate plan to escape—so elaborate that it sounded more like a mystery novel or thriller movie than something that could be pulled off in real life.

Harvey told me that he cruised local gay bars for three or four nights looking for a man who looked like him. When he found an appropriate candidate, he would say that he worked in a hospital and was concerned about AIDS, so before he would have sex, the other man would have to agree to let Donald test his blood. Once he had the blood sample, instead of testing it for HIV, he would type and cross-match it.

At the time he was living in a trailer heated with natural gas. He knew a funeral director who said he could get hold of dynamite, because it was sometimes used in cemetery digs with large rocks under the soil. Harvey's idea was to kill this individual who resembled him, place his body in the trailer, and then make it look like there had been a gas explosion.

I asked Harvey where he had planned to go. I said my best guess was Mexico. Strangely, that was the one thing he wouldn't reveal to me, as if it was part of an escape plan he was still harboring after all these years.

At any rate, the plan never went anywhere. The one man he found who he thought resembled him enough turned out to have the wrong blood type. And he indicated that he later realized that blood typing would not be enough—the scheme would have been uncovered by DNA analysis. But by then, presumably, Harvey would already be on the lam.

But what was interesting here was the degree of fantasizing that went into this escape plan, just as he was now fantasizing about building a house. My impression was that he needed these dreams to keep himself on an even emotional keel, just as Bundy had with his confidence in his own manipulative and evasive talents. In other words, there was always somewhere for Harvey to escape to, if not in real life then at least in his mind. This is consistent with the active fantasy lives of other serial predators.

The fantasy-type thinking pervaded many aspects of Donald Harvey's psyche. When I asked him why he had felt so depressed that he tried to commit suicide several times, he sloughed it off with "If I wanted to do that, I mean, I would have done it the proper way."

I came to believe that this need to keep everything on a simple and superficial level was inherent in his character. Going back to the day he skipped work so as not to show up for the polygraph examination that he and other hospital employees were supposed to undergo, he claimed he didn't remember the specific day, but offered, "I knew sociopaths could pass if they wanted to."

"Do you consider yourself a sociopath?" I asked.

"Well, that's what most people consider me," he responded. "What do you think?"

The benign, almost beatific smile came back. "I am just Donald Harvey. I haven't changed a bit. I have been rehabilitated. I am all right. I am ready to go back to the streets."

Perhaps nothing more clearly shows Harvey's inability or unwillingness to seriously confront his own psyche and motivations. Or else he's already done so but thinks the rest of us are too dense to catch on.

As the interview was concluding, Harvey essentially admitted as much: "I don't look at things, okay? For thirty-five years I was a free man; what I was doing I thought was right and the patients that I took care of I like to think, you know, I made their passing easy. They didn't give me permission—no. But some of the patients didn't have no one to give permission for them, just like I'm going to have a living will in case I get bad. We got good hospitals, but I am just saying, I don't want to live if I got to be a vegetable."

"Would you let someone treat you like yourself?"

"If they want to come, go ahead." Then he added, "Well, don't get caught, damn, don't get caught, because, you know, those little cells and bathrooms you live in ain't all that great."

His words turned out to be prophetic. On the afternoon of March 28, 2017, Harvey was found beaten unconscious in his cell at the Toledo Correctional Institution. The wounds were the result of blunt force trauma without a weapon, and the assailant was believed to be another inmate. Both killer and victim were said by prison spokesmen to have been in a protective custody unit. Harvey died two days later without regaining consciousness at Mercy Health–St. Vincent Medical Center in Toledo. He was sixty-four years of age—about thirty years short of his first possible parole date.

20

FALLEN ANGEL

Stepping back from the conversation with Donald Harvey, one can find it easy to be dismissive about the detached way he spoke about the killings and how they define his psyche. When Donald Harvey stated that he was "all right" and "a very warm and loving person," he was not just being flippant. While we certainly don't see him that way, he was raising profound moral and philosophical questions similar to those raised by the Nazi defendants at Nuremberg, particularly the concentration camp commandants and functionaries and those involved with running their systematic mechanization of degradation and death. Within their own logic system—a system devised by psychopaths and "normalized" by the state bureaucracy—what these subordinates were doing was not only acceptable but constructive. Either they really believed in what they were doing or they left all moral questions for the higher authority.

Similarly, Harvey certainly constructed his own logic system in which what he did was acceptable and beneficial—either as mercy killing or serving as an avenging angel. The institutions that employed him were either blind to what he was doing or unwilling to recognize the patterns of death that attended his patients and shifts. And unfortunately, this is not unusual. Institutions tend not to want to get involved in problems or legally hung up by delving too deeply. It is far easier just to pass the problem on to the next person or place down the line. How many hundreds or thousands of children were sexually abused because Catholic church officials considered it easier and less institutionally risky to reassign offending priests to other geographical locations or parishes than to confront and deal with their crimes?

Though Harvey's raw intelligence level is difficult to discern, he was highly resourceful and criminally sophisticated on an intellectual level. On an emotional level, he was just the opposite—a combination that made him a particularly dangerous and successful killer.

Unfortunately, he is far from unique.

At his sentencing hearing in Cincinnati in 1987, he stated, "There are several Donald Harveys out there."

He turned out to be right—retrospectively and prospectively.

In the first edition of the *Crime Classification Manual*, we listed hospital killings as a subset of Personal Cause Homicide. By the third edition, we had moved them out of Personal Cause and, due to their relative frequency, given them their own category: Medical Murders.

What Donald Harvey did is now subclassified as Pseudo-Mercy Homicide.

A man named Charles Edmund Cullen followed directly in Donald Harvey's footsteps. Except for the fact that Cullen was a divorced heterosexual with children, his case offers an uncanny parallel to Harvey's and graphically demonstrates the value of the offender interviews we do. Because of our understanding of Donald Harvey, we understand Charles Cullen and his like.

Cullen was an unassuming man who, like Donald Harvey, had had a traumatic childhood. Born in West Orange, New Jersey, in 1960, he was the youngest of eight children. His father, a fifty-eight-year-old bus driver, died when Charles was a baby. Like Harvey, Cullen attempted suicide several times, but didn't wait until he was an adult. His first try occurred when he was nine, drinking chemicals from a home chemistry set. When he was seventeen, his mother was killed in an automobile accident. His sister was driving the car. Like Harvey, Cullen enlisted in the military—in his case the navy—and received a medical discharge because of signs of mental instability and attempted suicide. And like Harvey, Cullen insinuated himself into a series of hospitals, studied their routines and procedures, and took advantage of their systemic failures of oversight.

In fact, the only major attribute Charles Cullen did not share with Donald Harvey was Harvey's superficial charm. Cullen was a withdrawn loner whom most people thought was odd as soon as they met him.

Over a sixteen-year period, Cullen worked as a nurse in nine hospitals in New Jersey and Pennsylvania, and during that time he killed at least thirty patients. There may have been ten times as many; no one, including Cullen, is certain.

Like Harvey, Cullen claimed to be an angel of mercy, putting people out of their suffering and misery. Also, as was true with Harvey, a good many of Cullen's patients were not terminal or were likely to recover. He showed the same muddled rationalization as Harvey did and seemed incapable of delving into his own motivations. And he blamed the hospitals for letting him get near patients whom he ultimately killed. His weapons of choice were insulin, epinephrine (adrenaline), and the heart medication digoxin.

Even more so than Harvey, Cullen aroused suspicion several times, each time affording hospital and law enforcement authorities the opportunity to end his deadly career. Several times other nurses noticed and reported unusual death rates on his shifts, and on several occasions, Cullen was released for substandard work performance. At St. Luke's Hospital in Bethlehem, Pennsylvania—by my count, the seventh hospital at which Cullen had worked—large quantities of rarely used drugs would disappear, and they would be replaced automatically. When a similar quantity would disappear again, it would be replaced routinely, and none of this seemed to arouse any suspicion or even prompt any questions. Cullen later said that his fingerprints—figuratively and literally—were all over these thefts.

But each time the authorities dropped the ball, concluding that there wasn't enough evidence or the situation was too ambiguous. Part of this was due to a national shortage of nurses, which allowed him to stay in jobs from which otherwise he would have been fired. Also, the hospitals just didn't want to acknowledge that there might be something going on within their wards beyond the normal death rates of people with life-challenging medical conditions. Even the hint of something like that could open a Pandora's box of uncertainties, charges, and financial exposure.

"How the Pennsylvania investigators didn't feel like there was enough evidence to prove it was me, to take away my license, I don't know," Cullen later commented to detectives. "I was the only person on all those nights, every single time it occurred."

Cullen's ultimate downfall came in 2003 at Somerset Medical Center in Somerville, New Jersey, where he'd been working for about a year in the critical care unit. By late spring, the hospital's computerized record system showed that Cullen was calling up the records of patients that were not his, and other personnel observed him repeatedly in the rooms of patients to which he was not assigned. Many of the pharmaceuticals were kept in locked cabinets controlled by an automated dispensing system known as Pyxis, requiring a computer access code to open. The record showed that Cullen was removing drugs that had not been prescribed for any of his patients. Often he would cancel the order quickly to try to cover his tracks, then order up the same prescription minutes later.

Around the same time, Dr. Steven Marcus, executive director of the New Jersey Poison Information and Education System, alerted the hospital administration to the possibility that four deaths were suspicious and could indicate an employee purposely killing patients. Hospital officials tried to diminish or moderate Dr. Marcus's concerns, but he held fast, telling them that he intended to report the case to the state Department of Health and Senior Services and that he had been taping the conversation. That got their attention, but they allowed Cullen to remain on the critical care ward until an investigation was mounted. Meanwhile, the suspicious deaths continued.

When a patient died in October with extremely low blood sugar—suggesting a possible insulin overdose—the administration finally informed law enforcement authorities. When the police began investigating, they uncovered a long trail of mistakes and oversights, including a nonfatal insulin overdose by Cullen in August. That was when his long record of hirings and firings by numerous hospitals in New Jersey and Pennsylvania came under scrutiny. Somerset Medical Center then took what many considered the easy way out—terminating Cullen for lying on his employment application. Meanwhile, the police investigation continued.

Cullen was arrested on December 12, 2003, while eating at a restaurant. He was charged with one count of murder and one count of attempted murder. Two days later, he agreed to be interviewed by Somerset County detectives Timothy Braun and Daniel Baldwin. And like Donald Harvey, when he got going, he gave the investigators a shock. During a seven-hour interrogation session, Cullen admitted to killing around forty hospital patients.

More than six hours into the questioning, Braun asked, "And the question that arises out of all of this, Charles, is the 'Why?' Could you please explain the 'Why?' behind all the deaths caused by yourself throughout the years."

"My intent was to decrease suffering in people I saw throughout my career," he responded, echoing Harvey. He went on to say that he had considered leaving nursing, because as long as he remained in the profession, "I knew that if I was placed in these situations that I would feel the need . . . to end suffering." He had to stay with it, though, because he had financial obligations, didn't want to be "a deadbeat dad," and knew he could not find another profession that would pay him as well as nursing.

He conceded that he "felt guilty for what I had done, even though I was trying to reduce people's suffering. I would go long periods of time with nothing, but then I would find myself getting back, feeling overwhelmed, feeling like I couldn't watch people hurt, and die, and be treated like nonhumans, and at times, the only thing that I could feel like I could do was to try and end their suffering, and I don't believe I had that right, but I did it anyway."

When we started studying serial killers back in the 1970s and '80s, we realized that most of them had what we referred to as a "cooling-off" period between crimes, whether that period was a few days, a few weeks, or even a few years. But then the internal pressure would build up again and they would be back to it.

Within just about every serial predator, there are two warring elements: a feeling of grandiosity, specialness, and entitlement, together with deep-seated feelings of inadequacy and powerlessness and a sense that they have not gotten the breaks in life that they should. And even though neither one is a predator in the traditional sense, we can see these traits clearly in Harvey and Cullen. Their pent-up feelings of powerlessness and disenfranchisement build up until they have to show their power—and in their twisted minds, their goodness and benevolence—by playing God with these unfortunate sufferers. The killing appeases their need for power and omnipotence, while at the same time satisfying their sub- or semiconscious need to strike back at the society that has denied them their due.

On March 2, 2006, Cullen was sentenced to eleven consecutive life terms, which makes him ineligible for parole in 397 years. He now resides at the New Jersey State Prison at Trenton, the institution that also houses Joseph McGowan.

Law enforcement is very much interested in crimes like Harvey's. Unfortunately, often we become involved only after the offender has made some blatant or inadvertent mistake. When we finally do get to examine such an offender, we get to comprehend the early disturbances and dysfunction in their lives that created the need for some form of retribution and retaliation. Donald Harvey was not going to allow anyone to treat him with disrespect ever again.

Studying medical murderers like Harvey, Cullen, or Shipman leads us to certain investigative considerations, but they do require a high degree of awareness, not only from law enforcement, but from people who are not normally trained to look for or recognize illegal activity such as drug or equipment theft or violent crime like murder. In any type of health-care setting, a statistically unusual rise in the number of deaths or unexpected medical complications should always arouse sufficient suspicion to warrant an investigation by the administration and then, if indicated, law enforcement. Is there any correlation with a particular individual's shifts or presence on the floor?

As we say in the *Crime Classification Manual*, "In nine cases of pseudo-mercy killers cited in an article in the *American Journal of Nursing*, the correlation between suspect presence and a high number of suspicious deaths was deemed sufficient to establish probable cause and to bring indictments by grand juries."

Another consideration would be an unusually high rate of cardiopulmonary resuscitation, either in the same patient or various patients. A particularly significant indicator would be if the same individual is present at many of these scenes or actually calls the emergency code.

If a suspect has had frequent job changes, that is certainly an element that should arouse suspicion. Unfortunately, as we've seen, administrators sometimes find it easier to simply let an employee go rather than follow up any possible leads against him or her.

We all accept that medicine is not an exact science, and so unforeseen outcomes will occur. But the most important investigative consideration for this type of crime is pattern recognition. The stakes are just too high for the security of one of our most vulnerable cohorts for us to fail in this area.

Because one thing my interview with Donald Harvey revealed was that he had no remorse and really liked getting away with murder.

IV

“NO ONE MADE ME DO ANYTHING”

21

THE SUPERBIKE MURDERS

Late in 2004, I was giving a lecture at a university in South Carolina. Afterward, Detective Sergeant Allen Wood of the Spartanburg County sheriff's office came up to me. He told me they had a year-old case in the town of Chesnee in which four people had been shot to death.

"Can you do anything to help us?" Wood asked.

"If you give me the information, I'll see if there is enough psychopathology evident for me to do anything with it," I replied, skeptical that I could help, since routine robberies often don't yield up much behavioral evidence.

When I returned home, Wood called to follow up and then sent me crime scene information and photos, autopsy protocols, and other reports.

On November 6, 2003, just after three in the afternoon, a customer and friend of the owner named Noel Lee came to the Superbike Motorsports motorcycle store and repair shop on the outskirts of Chesnee, a small farming community in the northwestern part of South Carolina. He saw a bloody scene with three dead bodies. He called 911:

"Where's your emergency?"

"It's at, uh, Superbike Motorsports. Apparently, everybody's been shot up here! Everybody's laying down in a pool of blood. His momma's been shot, the mechanic's been shot . . ."

After police arrived, the victims were identified as the store's owner, Scott Ponder, thirty years of age; service manager Brian Lucas, twenty-nine; mechanic Chris Sherbert, twenty-six; and Beverly Guy, fifty-two, Ponder's mother and the shop's part-time bookkeeper. All four had died of multiple gunshot wounds and there were eighteen shell casings of two different kinds littering the scene, both nickel and brass.

It appeared to investigators that the gunman or gunmen had entered the premises, gone to the shop area in back, shot Sherbert, then proceeded forward into the showroom, where he shot Guy. Lucas fell at the front doorway and Ponder in the parking lot, apparently trying to run for help when they saw Guy killed. There were no fingerprints or DNA evidence.

By definition, this would be classified as a mass murder. But the most significant indicator was what didn't happen: Nothing was taken, though there were thousands of dollars of cash in a briefcase prepared for bank deposit and many expensive and easily mobile motorcycles. So this went in my mind from criminal-enterprise homicide to some sort of violence in the workplace. And reconstructing the event from the evidence, it did seem that the mechanic was shot first, in the back of the head and from above, while he was working on a bike. He probably wasn't even aware of the shooter entering the shop area. The shooter then quickly went toward the front, where he encountered Guy coming out of the restroom and immediately shot her.

Lee told the sheriff's deputies he saw a young man and woman walking away as he entered the store. Kelly Sisk, who came forward when he heard about the slayings, says he was in the store about a half hour earlier with his four-year-old son to make a payment on a go-cart he was purchasing for the boy. He noted Scott Ponder assisting a customer who was looking at a black Kawasaki Katana 600 bike. Two things struck Sisk. The customer was wearing a black Columbia fleece jacket despite the fairly warm weather, and he didn't appear to have much experience with motorcycles. A black Kawasaki Katana 600 was in the shop area being prepared for delivery when detectives arrived. A bill of sale had been made out, but there was no name on it. Sisk gave a description of the customer, from which a sketch was made and circulated. As far as anyone could tell, Sisk was the last customer to leave Superbike before the shooting.

The sheriff's office followed up on a multitude of leads and theories. It could be a disgruntled employee or dissatisfied customer, or even someone hired by a competitor to put the successful enterprise out of business. Was the young man Lee saw the shooter and the woman the lookout? Lee himself also came under suspicion for having shown up alone just after the murders.

It was highly unlikely Beverly Guy was the main target, and nothing in Scott Ponder's or Brian Lucas's backgrounds aroused any suspicion. There were rumors of an illicit drug connection with Chris Sherbert, but nothing positive panned out.

Then an intriguing and damning piece of evidence turned up. The sheriff's office called Ponder's grieving widow, Melissa, and told her the baby boy to whom she'd given birth shortly after the murders was not Scott's. When she had been in the office on a previous visit and changed the baby's diaper, they had retrieved a DNA sample, and it did not match Scott's blood collected at the crime scene.

Melissa was shocked beyond belief and refused to concede the finding. She and Scott were completely in love. The only possible explanation to her was that they had given her the wrong baby at the hospital, but that seemed totally farfetched. Indignantly she demanded that a second test be conducted. The authorities complied. Not only did those results indicate the baby was not Scott's, but the child was fathered by his close friend and business partner Brian Lucas!

Was this a love triangle gone bad? There were rumors about the Lucases having marital problems, and shortly before the murders, Brian had been seen looking at a house on his own. Detectives were suspicious of Melissa because when they told her of Scott's death, she didn't want to hear the specific details. She later explained that she wanted to remember him as he was in his vibrant life.

I didn't know what to make of the DNA evidence regarding Scott's fatherhood, but after analyzing the materials sent to me, I didn't think any relationship between Brian and Melissa, if there even was one, had anything to do with the crime. Scott and Brian had both been killed and there was nothing in Scott's and Melissa's backgrounds to suggest they were having any problems; everyone who knew her affirmed Melissa was devastated by his death. Therefore, you just had to jump through too many logical hoops to make that scenario work. The sheriff's office had followed up on the drug possibilities and hadn't found anything, and there was nothing at the scene that suggested a drug hit to me.

Eighteen months after the murders, the sheriff's department was informed that vials containing Brian's and Scott's blood collected at the crime scene had been mislabeled. Melissa's baby had, indeed, been fathered by Scott, and at least one sorry chapter of the Superbike murder case was closed.

Was it possible a rival business venture had set up the murders? It wasn't impossible, but it was easy enough to check out, and from my experience, that is not the way legitimate enterprises conduct their affairs. It isn't even the way organized crime does it much anymore.

I gave Wood my assessment over the phone. He took notes and later shared them with me. Despite the two kinds of shell casings, this was the work of an individual. I said the killer was a disgruntled employee or customer, very probably a customer who was pissed off for some reason; an employee would be easier to trace. The sustained aggression of the multiple shots indicated that this was not a robbery that went awry, and the focus of the aggression was the entire company, rather than a specific person. Unlike most predatory, sexually based crimes, the UNSUB's age would not be a determinant, so it wasn't worth speculating about and potentially eliminating any suspects.

He would have spoken of his displeasure to one or more people, and at a certain juncture he just reached his boiling point. This was well planned and efficient. He would have been practicing his shooting beforehand, probably at a local range. He was scouting the location first to make sure there were no other customers and when Kelly Sisk left, saw that the coast was clear.

Post-offense, he would have been obsessed with the investigation and news coverage, with what I used to call a "high ass-pucker factor" in case the sheriff's office had any strong leads. Some of these guys actually come forward and inject themselves into the investigation to misguide the police or to come across as helpful. An example would be someone who says he was passing by and thought he saw a particular kind of car leaving the scene. This would not only deflect the investigation, but "explain" his presence if anyone happened to see him.

This long after the crime, with no arrest, the potential suspect's behavior would be back to normal. But I felt pretty certain that he had told several people something, perhaps bragging about his efficiency and how he got back at these people who had wronged him.

Though he would no longer be acting "strange," I suggested two proactive approaches. One was to go through the entire customer files to look for a complaint letter or to see if anything stood out or pointed to any leads. The second was to get a local news outlet to write a story and describe what the UNSUB's immediate post-offense behavior would have been like to see if anyone witnessed it or had been told by him.

After interviewing me, reporter Janet S. Spencer published a story in the *Spartanburg Herald-Journal* and *GoUpstate*, its online news service, headlined "Killer Profiled as Angry Gunman," and described my assessment.

"The number of times that Ponder and Lucas were shot as they fled indicated that the killer released the brunt of pent-up anger on them, Douglas theorizes," Spencer wrote. "He said robbery was not a motive. No money was missing from the cash register. No jewelry or personal belongings were unaccounted for on the bodies. Douglas said the crime does not even fit that of a drug-related mass killing . . .

"There is evidence that there have been disgruntled customers—either in previous dealings with the victims or the shop itself, according to the files, Douglas said. 'And the unknown subject may have had retaliation on his mind for months,' he said."

The article went on to describe pre- and post-offense behavior. " 'He has trained. Possibly going to local firing ranges if there are any, or just going out in the woods to target practice. He was accurate when he unloaded all those rounds at that shop,' Douglas said . . . He is likely a hothead who fights it out when he disagrees with someone. 'In this case he shot it out rather than talking,' Douglas said. The killer had no remorse, and that's why he was able to kill all four."

Despite more leads and investigators' ongoing efforts, the case remained stalled. And the killer, though, was still at large.

22

WHAT HAPPENED TO KALA AND CHARLIE?

Kala Victoria Brown, thirty, and Charles David Carver, thirty-two, were missing, and those who knew them well were terrified for their safety. The last time anyone had seen them was leaving the apartment they shared in Anderson, in the northwestern corner of South Carolina, on August 31, 2016. They had been dating for several months and their friends knew the relationship was serious. There were no text messages from either one after that date.

I knew none of this at the time. I learned some of it when follow-up stories began appearing in the newspaper. The rest I learned from what became a voluminous case file.

Carver was married to Nichole "Nikki" Nunes Carver, but they were in the process of getting a divorce.

Carver's mother, Joanne Shiflet, said she and her son never went a day without some sort of communication. She called the manager of the apartment complex where the couple lived. The manager went into the apartment and found no sign of them, only Brown's Pomeranian dog, Romeo, without food or water. Brown's mother, Bobbie Newsome, insisted that Kala would never voluntarily leave Romeo like that. And there was no sign of Carver's white Pontiac.

Posters with their photos went up and police entered the search. Some cryptic posts appeared on Carver's Facebook page saying they were okay and had just gone off by themselves, but Shiflet told investigators they didn't sound like her son. Someone may have hacked his account. And still, no one had actually heard from either Brown or Carver.

On October 18, Detective Sergeant Brandon Letterman of the Spartanburg County sheriff's office got a visit from two detectives from Anderson. They said they were working a missing persons case, and they had a tip that Kala was buried on a hundred-acre wooded property. Brown's cell phone had last pinged to a cell phone tower in Woodruff, just south of Spartanburg. The only property that fit the tip's description within two miles of the cell phone tower that had registered the ping belonged to a local successful forty-five-year-old real estate broker named Todd Christopher Kohlhepp, who lived in the Kingsley Park subdivision in Moore, southwest of Spartanburg, held a pilot's license, and owned a BMW sports car. Woodruff was about five or six miles farther south.

The sheriff's office flew a helicopter over Kohlhepp's property, looking for clues or evidence, such as Carver's car. But the dense forest revealed nothing. With a court order, Letterman obtained Kohlhepp's cell phone records, and when they arrived two weeks later, he found that the real estate agent's phone and Brown's had been in close contact at the time she disappeared. That was enough for a probable cause search warrant of both of Kohlhepp's properties.

On November 3, the sheriff's office dispatched two teams—one to Kohlhepp's home in Moore and the other to his Woodruff property.

Deep in the woods, the Woodruff team came upon a fifteen-by-thirty-foot green metal Conex shipping container, three-quarters of a mile from the nearest road. It was secured with five locks. The team worked with sledgehammers for fifteen minutes trying to break them.

Suddenly someone said, “Stop!” He thought he heard knocking from inside. Brandon Letterman knocked back.

He heard a faint “Help!” through the metal wall.

Using power tools they found in a barn on the property, including a blowtorch, deputies cut the locks and opened the door. They rushed in, guns drawn.

Inside the dark space they found Kala Brown, fully dressed and wearing glasses, but chained by the neck to the wall and handcuffed. “Just the girl! Just the girl!” the lead deputy called back as he surveyed the interior. “How are you, honey? These are bolt cutters and this is a paramedic. We’re going to get you out of there, okay?”

As they were cutting her loose, one of the deputies asked, “Do you know where your buddy is?”

“Charlie?” she said, still bewildered.

“Yes.”

“He shot him.”

“He shot him? Who did?”

“Todd Kohlhepp shot Charlie Carver three times in the chest.”

Letterman’s team relayed the information to the Moore team. At the house, senior investigator Tom Clark, accompanied by Anderson investigator Charlynn Ezell and senior investigator Mark Gaddy, confronted the 300-pound, disheveled-looking Todd Kohlhepp with what they had learned. Kohlhepp asked for a lawyer and to speak to his mother. He was taken in handcuffs to the Spartanburg detention center, where both requests were granted.

Meanwhile, the Woodruff team had searched the loft apartment above his garage and found chains and shackles. “You don’t see that too often,” one of the deputies commented.

Deputies found Carver’s car, stained brown to help hide it, on Kohlhepp’s property, crushed under tree branches, covered with a pile of brush. They also found a prepared but empty grave.

In the ambulance that took her to the hospital for a checkup, Brown revealed to detectives that Kohlhepp had told her he was responsible for multiple murders in a motorcycle shop "a few years back." It had actually been thirteen years.

More surprises were in store when, at the beginning of a voluntary four-hour-long confession, Kohlhepp announced, "I'm going to close a few cases for you." He identified the Beretta handgun and the types of ammunition used in the Superbike murders, details that had never been released to the public. Then he went on to describe how he had killed Johnny Joe Coxie, twenty-nine, and Meagan Leigh McCraw Coxie, twenty-six, local residents who were reported missing on December 22, 2015. Kohlhepp said he had hired the couple to clean some of his rental properties and had stopped off at his place in Woodruff to pick up supplies. But when they got there, he thought they were trying to rob him when Johnny pulled out a knife. A few days after the interrogation, he brought investigators to the spot on his property where he had buried them. He said he had shot Johnny right away and kept Meagan alive for several days trying to figure out what to do with her, before finally deciding that killing her was his only option.

Throughout all the questioning, detectives reported that Kohlhepp was calm, patient, and matter-of-fact, expressing neither remorse nor regret that he had been caught, though he must have felt some of that. The only emotion he showed was occasional pride in his prowess.

"I cleared that building in under thirty seconds," he declared, sitting across from two detectives in the interrogation room. "You guys would have been proud. My golf game is weak; my kill game is strong."

He also said he and some other men had killed drug dealers in Juárez, Mexico, on several "hunting trips."

Kala's revelation was not the first time the sheriff's office had come across Todd Kohlhepp's name, however. As part of the Superbike investigation, a form letter had been sent out to everyone on the motorcycle shop's customer list asking recipients to contact them if they had any information about what had happened or that might lead to an arrest. Not surprisingly, he never responded. But the investigators did not go so far as to interview each of the hundreds of names on the list.

If they had, his name almost certainly would have stood out, not because of anything having to do with his purchase of a motorcycle—though his attempt to return it might have presented a clue—but because Todd Kohlhepp was a registered sex offender. And that should have aroused enough interest at least to bring him in for an interview.

This is often the way crimes get solved—through a side door. The clue that led to the closing of New York's "Son of Sam" killings was a parking ticket David Berkowitz was issued for parking his Ford Galaxie too close to a fire hydrant near the site of his final murder.

Despite the horror of the seven murders attributed to Kohlhepp, the most fascinating and salacious aspect for the media was that he had apparently kept a young woman chained up as a captive and sex slave for more than two months. From a hospital room where she was taken to be examined and recover, and later in a series of interviews, Kala Brown recounted how on the morning of August 31, she and Charlie went to Kohlhepp's house for a job he had hired her for. Charlie came along to help her. She had done other cleaning jobs for Kohlhepp and his real estate firm after posting on social media that she was looking for work. He hired her repeatedly, though he complained to police during his interrogation that it had taken her three days to do a job that should have taken one.

When I first heard about Kala Brown's rescue, I assumed that she was the victim of a sexual sadist for whom the imprisonment, degradation, torture, and rape was the signature aspect of the crime. I thought immediately of Gary Heidnik, who imprisoned, raped, and abused women in the basement of his house in Philadelphia, whom my colleague Jud Ray and I had interviewed in the penitentiary in Pennsylvania. But I came to realize I had jumped to a conclusion based on appearances.

Kohlhepp told Kala and Charlie he would lead them to his wooded land in his car because he had to unlock the gate to his property, which he called a farm, about a fifteen-minute drive from the house in Moore. At the entrance to the property, he got out and unlocked the metal gate, then locked it again after Charlie's car had driven through.

They followed his car about half a mile or so down a road, past fields and trees, until they came to a clearing with a large two-story garage with a barn-style roof, a small garden shed, and a metal storage container. They went inside the garage, where Kohlhepp handed them each a pair of hedge clippers and a bottle of water. He said they would be clearing underbrush from trails and he would show them where to start. They came back outside, but he said he had to go back inside to get something. Kala and Charlie stood outside for several minutes, holding hands as they waited.

Kohlhepp claimed he heard them talking about stealing from him, so when he came back out, he shot Charlie three times in the chest. As he later told investigators who brought him out to the scene, he got his Glock .22, "and then I came out and dropped him right about here."

Kala said that as she stood there in stunned disbelief, he pulled her forcibly inside the garage, saying that if she didn't go willingly, she would join Charlie. He handcuffed her behind her back, ankle-cuffed her, and put a ball gag in her mouth. He said he had to go take care of Charlie. He was completely calm.

He brought her back out about twenty minutes later. Charlie's body was lying in the front loader bucket of a tractor, wrapped in a blue tarp. He told Kala he had held another woman sometime back, but at some point, she "pissed him off" so he killed her with a shot in the back of her head. He said he had committed many other murders, close to a hundred, some while he was in prison and was let out by the government from time to time to serve as a foreign hit man.

For the first two weeks of her captivity, he kept her chained to the wall of the Conex container, bringing her into the larger building twice a day to eat and have her do "whatever he wanted sexually." If she refused his sexual advances, he wouldn't force himself on her. "But he made it very clear why I was there, and if I wasn't useful, then I wouldn't need to be kept any longer, and then he would shoot me. He said if I was a good girl, he'd teach me how to kill and I'd get to be his partner."

As the time of her captivity wore on, she was kept mostly in the container, often in the dark, and led to the house for some meals and to use the bathroom. She said she tried to be cooperative so he would treat her better.

Sheriff Chuck Wright announced that Kala Brown would receive the $25,000 reward, long offered for information leading to the arrest and conviction of the perpetrator of the Superbike murders.

During his interrogations, Kohlhepp admitted what he had done to and with Brown, but he offered a somewhat different perspective. He never beat her, never physically harmed her, and the sex was consensual, he claimed, at her instigation. He said she had a long list of material requests, which he dutifully ordered for her on Amazon to keep her contented. Though she said she believed he was infatuated with her, he claimed the reason he kept her imprisoned in the Conex container, he said, was that he couldn't figure out what to do with her once he had impulsively killed Charlie Carver.

He said she was a drug addict and that he "got her clean" while he held her. He said he had "a real hard time with drug dealers" and resented that she was using the money he paid her for drugs.

Barry Barnette, the Seventh Judicial Circuit solicitor, conferred with families of the victims before offering a plea deal, and said he would abide by their collective decision. Most realized that it could be decades or more before an execution was carried out. Scott Ponder Jr., who never got the chance to know his father, agreed with his mom, Melissa, about the proposed arrangement.

On May 26, 2017, in exchange for avoiding a trial that could have resulted in the death penalty, Todd Kohlhepp pled guilty to seven counts of murder, two counts of kidnapping, one count of criminal sexual assault, and four counts of possession of a weapon during the commission of a violent crime. He was sentenced to seven consecutive life sentences without the possibility of parole, plus sixty years.

23

WHAT MADE TODD TICK?

When I read about Todd Kohlhepp's arrest, I was as gratified as I always am when a killer has been caught. I was also gratified that my profile of the Superbike UNSUB had been accurate, but sorry investigators hadn't gone through the entire customer list and checked out each one. I didn't think, though, that I would have anything more to do with the case.

Maria Awes is a documentary film producer who spent a decade in broadcast journalism, where she won awards for her investigative reporting. Together with her producer-director husband Andy, she formed Committee Films in Eden Prairie, Minnesota, a suburb of Minneapolis. One afternoon in 2016, she was meeting with Stephen Garrett, one of her associate producers, when he got a text from his cousin Gary in Spartanburg, South Carolina.

Gary Garrett was a real estate agent, and the text to Stephen was shocking. His former boss, Todd Kohlhepp, had just been arrested and charged with seven murders, and wanted Gary to write his life story. Kohlhepp was saying that 90 percent of the "real story" had not come out. Stephen knew Maria had a background in investigative journalism and thought she might be able to give his cousin some media advice.

Maria had read about Kala Brown's rescue in the news, as I had. She spoke with Gary by phone and told him what to consider and look out for if he was going to deal with an accused killer. And Kohlhepp's claim about the real story roused her reporter's instinct. "Do you think Kohlhepp would want to share his story with us?" she asked Gary. He said he didn't know, at which point she suggested he see if Kohlhepp would speak with her.

Shortly thereafter, Maria spoke by phone to Kohlhepp in the Spartanburg county jail. They were limited by regulation to two fifteen-minute calls, which Maria recorded. "I was struck by how he spoke: 'Yes, ma'am, no ma'am.' Very much like a southern gentleman," she recalls. "He told me the body count was considerably higher than what he had been charged with. He spoke freely, very calmly. I just kept asking questions. He said he felt bad about killing Charlie Carver and insisted that he never raped Karla, that all sexual relations between them were consensual.

"I said I wanted to know what this 90 percent untold business was all about. Kohlhepp agreed to talk and said he intended to plead guilty. I needed to see if I could get a network involved. Crime coverage was part of my investigative background, and I had always been interested in what makes someone commit murder. Why does he act differently than other people? We had been trying to get something going with Investigation Discovery, and this seemed like the perfect subject."

Maria talked to a producer at the Investigation Discovery network, who agreed to fund research for the film project. Maria went down to South Carolina and was able to talk to Kohlhepp via videoconferencing in the county detention center. The six-episode television series that emerged is titled *Serial Killer: The Devil Unchained*. The title refers to a pre-sentencing report from his teenage rape conviction in which a neighbor described him as "a devil on a chain."

As she got deeper into her research, Maria continued talking with

Kohlhepp. "After several conversations," she said, "I decided I needed to talk to someone who had done this before—talked to accused and admitted killers."

Which is when she contacted me, and after some discussion, we agreed that I would interview Kohlhepp. With her reporter's tenacity, Maria used Freedom of Information (FOI) requests to obtain the case files, which her talented research librarian Jen Blanck had meticulously organized and sent to me in voluminous binders and folders.

By this time, Kohlhepp had pled guilty, had been sentenced, and was incarcerated in the Broad River Correctional Institution in Columbia, South Carolina, where rules were tight and access to the prisoner was much stricter than when he had been held in the local jail. And frankly, after what he had admitted to doing, prison officials were not inclined to grant him any special favors. In fact, due to his local notoriety, he was kept isolated from the general prison population for a considerable period of time.

But Maria continued her communication with Kohlhepp by letters and emails. Along with Garrett, she was able to get information that Kohlhepp hadn't given the detectives.

His first gun crime, he told Maria, was in Arizona when he was barely a teenager. He wrote:

> *Yes, I shot someone in Arizona, not a drug dealer, some jackass that wanted to be in a gang and I was random part of initiation, unknown to me. He shot a friend of mine. I later emptied a gun into his car in parking lot one night. No clue what happened to him or if he was even hit. I was fourteen. I was young and scared, I know I shot out the window, but when gun was empty, I raced out of there and threw gun in dumpster in an alley.*

He told her he had once killed two thugs who tried to accost him in the parking lot of the Hunt Club apartment complex where he used to live.

The way he described it, two men, one hefty and the other smaller, confronted him. The smaller man stepped forward first, brandishing a knife. The other carried a hammer. Kohlhepp dropped his keys and grabbed two knives he kept in his pockets, one in each hand. The attacker holding the knife extended his arm and Kohlhepp sliced his wrist, causing him to drop his weapon. He tried to kick Kohlhepp, but Kohlhepp cut him on the inside of his thigh and then stabbed him in the chest. The larger man with the hammer lost his nerve and turned to flee, but Kohlhepp caught him from behind, grabbed him by the hair, and stabbed him in the side of his neck.

Kohlhepp went into his apartment to grab towels, blankets, and a shower curtain, wrapped the smaller guy, and hefted him into the trunk of his Acura Legend. He lowered the rear seat, spread out the shower curtain, partially wrapped up the larger guy, and put a towel over his face. He went back inside, found the largest pot in his kitchen, and filled it with water, making about a dozen trips back and forth trying to wash the blood off the parking lot surface. Then he drove off and found a closed road, beyond which he buried the bodies.

"I put the bodies behind the barricade in a ravine," he told Maria. "I'm surprised no one ever found them."

These bodies have never been located.

For my part, I was intrigued by the prospect of interviewing Kohlhepp, because he didn't appear to fit into the traditional categories we can assign to most repeat killers. I studied the transcripts of the arrest and interrogation that Maria had obtained under FOI, as well as what he had told her, such as these Hunt Club murders. Then I started putting the facts together. Todd Kohlhepp was a successful real estate broker who had other agents working for him. He was an accomplished pilot. None of his crimes would be classified as criminal enterprise—that is, for personal gain. All of his money was made legitimately. There seemed to be some sexual component to some, if not all, of his crimes, but I really had no idea what it was or if it was the primary motivator. Apparently, he hated drug dealers, which was a separate issue, but was willing to work with anyone who could get him the weapons he couldn't legally purchase on his own. After initially denying everything, as most criminals would, he was candid and forthright with his interrogators. And other than the occasional bragging about his shooting prowess, he was as calm and dispassionate in those sessions as Kala Brown had described him being after he shot Charlie Carver.

The Superbike murders were highly organized. The Coxie and Carver murders offered a mixed presentation, with organized and disorganized elements. He described how he killed Carver but couldn't really explain during the interrogation why he killed him. He had sex repeatedly with a woman he had imprisoned in a dark shipping container, yet he claimed to have backed off when she protested, and frankly, his actions during this time did not fit any of the established rapist typologies. He conceded that what he had done was wrong and didn't blame anyone else for his actions.

So what made Todd Kohlhepp tick?

On the outside, Kohlhepp had lived as successful and productive a life as just about any multiple murderer I could think of. He had a B.S. in computer science from Central Arizona College. He had worked for more than a year as a graphic designer. He passed the tests to receive a private pilot's license from the Federal Aviation Administration. He had passed the South Carolina real estate exam and had received a broker's license for the firm that he operated out of his house in Moore. He even wrote online reviews for products he'd purchased on Amazon.

Yet as we looked at some of those online reviews, another side of his personality emerged. For a chainsaw he wrote: "Works excellent. Getting the neighbor to stand still while you chase him with it is hard enough without having a easy to use chainsaw."

A knife received: "Haven't stabbed anyone yet . . . yet . . . but I am keeping the dream alive and when I do, it will be with a quality tool like this."

A folding shovel: "Keep in car for when you have to hide the bodies and you left the full size shovel at home."

A hidden shackle padlock: "Works great. Also, if someone talks back, go old school on them by putting this in a sock and beating them. They will not appreciate the hardened steel like you will. Works great on shipping containers."

KOHLHEPP WAS BORN TODD CHRISTOPHER SAMPSELL IN FORT LAUDERDALE, FLORIDA, on March 7, 1971. His parents, Regina and William, divorced when he was two. Regina got custody and soon remarried, to a man named Carl Kohlhepp, who had two children of his own and adopted Todd when he was five. Though possessing an above-average intelligence, Todd was by nature a difficult, often angry, aggressive, and rebellious child, and he clashed continuously with his stepfather. There is evidence of cruelty to animals and hostility to other children. At the age of nine, while the family was living in Georgia, he was sent to a state mental health facility for three and a half months for anger management. The Kohlhepps moved to South Carolina, where Todd was kicked out of the Boy Scouts for being disruptive. He wanted to go live with his biological father, whom he really didn't know, figuring he'd have a better life, and threatened suicide if he wasn't allowed to go. In desperation, Regina finally agreed while she and Carl were having marital conflict (they would divorce and remarry and divorce again) and he went to live with Sampsell in Tempe, Arizona, where he owned a restaurant called Billy's Famous for Ribs.

It wasn't long before Todd grew disenchanted with his father, who he said was always out with girlfriends and paid little attention to him. He told Regina he wanted to come back, but she found excuses to keep him with her ex-husband.

While Todd was living with his father, his behavior continued to deteriorate, culminating in an Arizona kidnapping conviction in 1986. The fifteen-year-old Kohlhepp got his father's .22-caliber handgun and went over to his fourteen-year-old neighbor's house, where she was babysitting her younger brother and sister. Todd forced her to walk with him back to his house, where he brought her to his bedroom on the main level, put duct tape on her mouth, bound her hands behind her back, and then raped her. She wasn't his girlfriend, though he wanted her to be. In fact, she was interested in another boy at school. He had tried to get her to come over to his house four times before, and she had refused, before he finally had the idea to force her at gunpoint.

After the attack, there is discrepancy in Todd's and the girl's versions of what happened. Todd said she willingly agreed to help him look for his dog, which had run outside, before he took her home. He did admit that he threatened to kill her younger siblings if she told anyone what he'd done. She said he was nervously pacing the floor, debating in his own mind whether to kill her, and it was her idea to concoct a story about looking for his dog to explain her absence from her house, promising Todd that if he let her go, this was what she would tell her parents.

Before she got home, though, her five-year-old brother had noticed she wasn't there and became alarmed. He had recently been taught how to dial 911. By the time the parents returned home, the police were already there. When the girl got home a short time later, she started out with the story about the dog, but then broke down and detailed the rape.

The police went to Todd's house, where they found him holding one of his father's rifles, pointed at the ceiling. When a probation officer asked him why he had assaulted the girl, he replied that he wasn't sure, but it could have been an act of rebellion because his father was out of town. He also said he thought the girl was sixteen rather than fourteen. He later said he just wanted to talk to her, to convince her to be his girlfriend, and the "situation got out of hand."

According to an article by Tim Smith in the *Greenville News*, "The parents of the girl told the officer that the rape had a 'devastating effect on the entire family.' The girl cried and was unable to communicate during most of an interview with the probation officer and her parents said her grades and athletic ventures had deteriorated."

This truth is in stark contrast to the version of the crime that Kohlhepp offered in order to obtain his real estate license in South Carolina twenty years later. In a 2006 letter to the South Carolina Department of Labor, Licensing, and Regulation, he explained that he and his then-girlfriend, both fifteen at the time, got into an argument at his house while his father was away. He had foolishly picked up his father's handgun, which he had out because he was afraid of robbery while he stayed in the house alone, and told her not to move while they talked out their differences. According to his account, her parents became worried when they couldn't reach her by phone and called the police, who showed up at the house. The lie worked, and the real estate license was awarded.

The same could not be said for the conviction of the crime itself. Todd was charged with kidnapping, sexual assault, and committing a dangerous crime against a child. The probation report cited a neighbor's comment that he was starved for affection and attention, but recommended he be charged as an adult. He agreed to plead guilty to kidnapping in exchange for the other charges being dropped. He was sentenced to fifteen years in prison and registered as a sex offender.

While in prison, he received his degree in computer science. When he was released in August 2001 after serving fourteen years, he moved to the Spartanburg area, where his mother was living, where he got a job as a graphic designer, a job he kept until November 2003, the same month as the Superbike murders. He enrolled in Greenville Technical College in 2003, then transferred to the University of South Carolina Upstate and was awarded a B.S. in business administration and marketing in 2007. By then he was already set up in the real estate business.

Initially, Kohlhepp seemed to be on a new track, and his coworkers' accounts support that perception as well. According to Gary Garrett, Todd had been a good boss, dedicated to his real estate business, driven and highly focused on marketing. He was very aggressive on behalf of his clients. He treated his agents well and very few complaints had been filed against him with the local real estate board. An article in the *Greenville News* stated, "Mortgage lenders described him as 'an effective communicator and a pleasure to talk with' and 'on it when it comes to getting a deal done for his clients.' One builder described him as 'incredibly personable.' "

But then something seemed to change in him. Gary said he went from "normal" to narcissistic and belligerent. He started bragging about his guns. There were more complaints filed against him. And his weight increased noticeably. This turned out to be shortly after Meagan and Johnny Coxie disappeared at the end of 2015.

There is always a motive for murder, even if it is not readily apparent or understandable; even if it is something as elemental as the power and sadistic stimulation someone like Dennis Rader felt from being able to watch a victim die and observe her fear and anguish. But Todd Kohlhepp was not that kind of killer. Whenever he killed, he had a more "logical" and "practical" reason.

The Carver murder was the one that most perplexed me. He told Maria Awes that he had overheard Charlie and Kala talking about robbing him and using the money to buy drugs to support her habit. We knew that drug dealers and perceiving that he was being taken advantage of were two of his hot buttons, so this made sense, but it was also very close to the explanation he used for killing the Coxies. Was he just paranoid, or was this merely an excuse for getting rid of the men so he could possess and control the women?

During his initial interrogation at the Spartanburg County detention center following his arrest, he said he had not prepared the Conex container as a prison. Rather, it "was designed for my food and my weapons and to secure my four-wheel before I had the [garage] building built." He said he had to clean out the area before he could put Meagan Coxie in there after he shot Johnny. "For the first time, I was having a little panic of what the hell to do with her—put her here, put her there, drop her, what the hell do I do? Do I call the cops? Oh shit, I got illegal guns. Oh shit, oh shit, oh shit! What do I do with her?"

The seemingly messy and haphazard nature of Carver's murder and Coxie's kidnapping provided a stark contrast to the Superbike killings, thirteen years earlier, which were a completely different type of crime. From the facts of the case and what Kohlhepp related to investigators, the Superbike murders would be classified in the *Crime Classification Manual* as "mass murder and personal cause: revenge and retaliation homicides." It intrigued me that the same individual would commit both.

We would have to classify Kohlhepp as a predator because of the way he planned the Superbike murders and the way he said he killed drug dealers in Juárez. But what fascinated me was that he was unlike just about any other violent predator I had encountered. He did not seek potential victims; in most cases they presented themselves to him. Yet he did not target victims of opportunity, either. Rather, he killed people in response to wrongs done to him, real or imagined.

I was also intrigued by the fact that the Superbike murders, the Juárez murders (if true), and the Coxie and Carver murders represented such a range of circumstances. Unlike the serial killers I had studied and even the single-victim killers like Joseph McGowan, this guy didn't fit into any apparent pattern. I wanted to know more about him.

Kohlhepp related to sheriff's office detectives that in 2003 he had bought a motorcycle from Superbike Motorsports—a Suzuki GSX-R750—for $9,000. He didn't really know how to ride, and it didn't really work out for him as he practiced. He went back to the store. "I thought [buying the Suzuki] was a bad decision. I was trying to see if I could possibly trade it in for a smaller bike or something of that nature." But he said they were "a little on the rude side about it, uh, my inability to ride that kind of bike," and he perceived that they were ridiculing him.

Three days later, he said, the motorcycle was stolen, and since

Superbike had delivered it, he believed someone from there had stolen it. To compound his indignation, he says that when he contacted the police to report the theft, "the law enforcement officer made fun of me."

He continued going back to Superbike, sitting on different models and imagining riding them. He listened as the manager and owner would "basically talk-trash" to each other. He bought a handgun—a Beretta 92FS—which he had to obtain illegally through a third person because of his sex offense registration.

On November 6, 2003, after checking out the store to make sure all the customers had left, he went in again. He went over to a black Kawasaki Katana 600, sat on it, acting as if he was feeling it out, then announced he would buy it. The mechanic wheeled it back into the shop to prep it. Kohlhepp waited a few moments, put on two pairs of latex gloves, then walked into the shop area, shot the mechanic twice at a downward angle, then proceeded with the rest of his controlled rampage, stopping once to reload, which could account for the brass and nickel rounds, though evidence at the scene suggests the otherwise finicky Kohlhepp mixed bullets in his clips.

He said it had "the desired effect."

He walked out and left the scene in his Acura Legend. At home, he took the gun apart, put the pieces in kitty litter, and then discarded them in several trash cans and a Dumpster.

He acknowledged to the detectives that he knew he would spend the rest of his life behind bars. His only concern, he said, was figuring out a way to leave his money and assets to his mother and his longtime girlfriend, to pay for her daughter's education.

24

"GOOD OR BAD, I STILL WANT TO KNOW"

Even though I was on board, it took a while before Todd Kohlhepp committed to an interview with me. In an email to Maria after I had written to him and introduced myself, he wrote, "I am unsure of John Douglas. My understanding of him is glamour hog that spins more than just puts it out what happened."

Once he had read our most recent book, *Law & Disorder*, which Maria had sent him, and possibly in response to the respectful relationship he'd built up with her, he wrote:

> *I wasn't very amused with John Douglas in his Law and Disorder book where he was extremely mean in his comments about women who write to or have any relations with inmates, especially death row, that they were all pathetic. Considering those are also the same people buying his books, I found he enjoys making cruel comments. I don't think we would get along at all, but I do respect his history and experience. I will agree to meet with him . . . I will agree to be open with*

a profiler only if they agree to explain [their] findings to me. Good or bad, I still want to know.

With the exception of Ed Kemper, who had figured himself out pretty well, I couldn't imagine other repeat killers in my experience being this genuinely interested in finding out why they behaved the way they did. It presented a rare opportunity.

What he was referring to as my "extremely mean" comments was my perspective that most women who fall in love with incarcerated killers were "pretty pathetic" and I felt sorry for them. We were not talking about professionals like Maria. And ironically, in that section of the book, we were describing a woman who did not fit that stereotype at all. In March 2006, a woman named Lorri Davis called and asked me to join the defense team trying to get a new trial and exoneration for her husband, Damien Echols, and two codefendants who had been convicted of the 1993 murder of three eight-year-old boys in West Memphis, Arkansas. The case had already gained a fair amount of notoriety as the result of two HBO documentaries, *Paradise Lost: The Child Murders at Robin Hood Hills* and *Paradise Lost 2: Revelations.* Damien, Jason Baldwin, and Jessie Misskelley Jr. had become known as the West Memphis Three. Damien, the supposed ringleader, had been on death row since his conviction as an eighteen-year-old in 1994.

Lorri, a successful landscape architect in New York, had seen the films, become interested in Damien's case, started writing to him, and eventually fell in love and moved to Arkansas to be near him and advocate for his innocence. It so happens that Lori and Damien are two highly intelligent, sensitive, and loving individuals and she convinced me to join the West Memphis Three defense effort, which, I soon found out, had been spearheaded and largely financed by New Zealand film director Peter Jackson and his producing and life partner, Fran Walsh. I accepted the invitation to join the investigation but gave Davis the usual warning—to be passed on to Jackson and Walsh and the other supporters—that there were no guarantees my analysis would help the appeals process because it would be evidence, not theories or advocacy, that would drive my assessment of the case.

After reviewing the voluminous case file, I concluded first that, contrary to the prosecution's entire case, this was not a satanic ritual murder. The murders occurred at a time when "Satanic Panic" had gripped the nation as the latest boogeyman trend and local police departments were even hiring self-proclaimed "experts" to help them solve these crimes. My other observation was that there was not a scintilla of evidence connecting Echols, Baldwin, or Misskelley to the murders, despite Misskelley's coerced confession from detectives of the West Memphis police department.

My analysis concluded that these killings were not the work of strangers, but rather personal cause homicides. Forensic and behavioral evidence at the scene pointed to someone who was criminally experienced and, in all probability, lived close to the three victims. The most likely suspect, who had a history of violence, had never even been interviewed by the investigators. I was able to convince the mother of one of the boys and the stepfather of another, both of whom had been certain the three teenagers had killed their children, that they had nothing to do with the crimes.

At the conclusion of our investigation, I participated in a press conference in Little Rock at the University of Arkansas School of Law, organized by Echols's appellate attorney Dennis Riordan, where various experts presented their findings. The other participants were Dr. Werner Spitz, the distinguished forensic and anatomic pathologist who literally wrote the standard textbook on medico-legal death investigation; Dr. Richard Souviron, the chief forensic dentist for the Miami-Dade medical examiner's office and an expert on bite marks; and Thomas Fedor, a criminalist, DNA expert, and blood and body fluid analyst.

Ultimately, the defense effort fell short of the exoneration we all felt was due, but the current district attorney agreed to an Alford plea, a legal technicality in which the defendants technically plead guilty while proclaiming their innocence. In return, they were let out of prison for the eighteen years' time served. In a very real sense, I considered this a deal with the devil, since no prosecutor or attorney general who actually thought the defendants had viciously killed three young boys would have ever let them out of prison. Instead, in my opinion, it was a cynical ploy to avoid a wrongful imprisonment lawsuit that likely would have cost the state of Arkansas tens of millions of dollars. We could have asked for a new trial, but that would have kept Damien, who was already in failing health due to his treatment on death row, behind bars for another several years, and we were worried he couldn't survive that.

AS IT TURNED OUT, KOHLHEPP WAS NOT THE GREATEST IMPEDIMENT TO A FACE-TO-face meeting. He was not a popular inmate, and the prison administration considered him a troublemaker and disruptive influence. Therefore, the warden and his staff were not going to give him any access to the outside world that they didn't have to. I went to the head of the Department of Corrections and appealed through a colleague at SLED—the South Carolina Law Enforcement Division, whom I'd trained in profiling when I was still in the bureau. But we kept hitting the proverbial brick wall.

Kohlhepp himself seemed disappointed. As he wrote to Maria:

> *Not being able to interview me in person is going to make this a bit more difficult, but not impossible. I spend a lot of time now thinking of my actions, why, what led to them and due to the high pressure environment I stayed in, what I thought led to them at the time. Not always the same looking back now that I don't have* [a] *hundred phone calls a day to interrupt.*

That gave me an idea. What if we could get Kohlhepp to fill out the assessment protocol we had used in our original serial killer study?

It had always been executed by the interviewers rather than the incarcerated offenders, but with someone as intelligent and expressive as Kohlhepp, it might be a highly effective alternative to a live prison interview. It would give the subject time to think about his responses, and with what we knew about the crimes, if he was lying, dissembling, or holding back, we would know it. And together with the questions he had already answered in his correspondence with Maria, I thought we could gather a complete behavioral portrait of Todd Kohlhepp and a sense of what made him tick.

I told Maria about protocol. She was immediately interested and agreed to mail it to Kohlhepp, with an explanation of what it was. Then we waited to see if he would cooperate.

He did, and with a completeness that surprised us. He is the only convicted killer ever to complete the assessment protocol himself, giving us direct, unfiltered access to his mind and the way he sees himself. And not only did he complete it, in numerous places he felt the need to go *beyond* the printed form, writing detailed explanations and narratives on many additional sheets of lined paper. I am confident the responses he gave were what he would have said to me face-to-face in the prison.

This approach wouldn't have worked with most offenders, but I was optimistic with Kohlhepp because as far as I could tell, he was introspective and above average in intelligence, and from the communications with Maria, he seemed as if he genuinely wanted to understand himself. The form couldn't have been used with David Berkowitz, Charles Manson, or Dennis Rader, say, who were too bound up in their own images to answer truthfully without someone like me staring at them across the table to break through their long-cultivated facades. The only other offender that it might have worked with was Ed Kemper, who was also pretty introspective and self-analytical, had we fully developed the assessment protocol at that point.

Here, with Kohlhepp, in addition to the extensive case files, I could triangulate among three sources—the police interrogations, the numerous letters between Kohlhepp and Maria, and the completed assessment protocol. Each of these three sources approached the questioning in a different way. The interrogations were adversarial; the letters from Maria were friendly, supportive, and inquisitive in tone; and the assessment protocol was neutral and objective. If Kohlhepp responded differently in each instance, that would be an immediate tip-off as to his trustworthiness. If his responses were consistent from one source to another, that would tell me something, too.

Practically speaking, he seemed to be offering us a trade. He would fill out the protocol document if I would give him my assessment of why he was the way he was and did the things he did.

Kohlhepp wasn't the first killer who wanted to understand himself through profiling techniques, though he was probably the sincerest about it, as opposed to merely using them as a stimulant to his own narcissism. The night before I interviewed Dennis Rader in the El Dorado Correctional Facility in Oswego, Kansas, I met at the cocktail lounge in my hotel with Kris Casarona, a woman who had established a relationship with Rader after his incarceration with the intention of writing a book and had become a sort of unofficial conduit to him. Both she and Kenneth Landwehr, the Wichita police homicide detective who pursued BTK and finally arrested him, had told me that Rader was a fan of the books I'd written with Mark, particularly *Obsession*, which had opened with a somewhat disguised version of the BTK case. The chapter was entitled "Motivation X," a takeoff on the Factor X that BTK had referred to in his letters to the police and media as the neuropsychological reason for his deadly predilections. Our chapter was written in the first person, as if from the killer's point of view, and published years before Rader was identified and arrested. He told Casarona that he had read the chapter over and over again and it had given him a sense of perspective and an understanding of the forces that swirled around his brain.

At the hotel meeting, Casarona handed me five handwritten pages from a yellow legal pad in Rader's tight, small handwriting that he had asked her to give me. They turned out to be Rader's own notes and evaluation of himself based on the traits we'd presented in the book. He had mailed it to Casarona a few days before. Across the top of the first page he'd written: "OBSESSION (CASE STUDIES)."

Rader had listed the attributes we'd mentioned as belonging to our serial killer UNSUB, together with the pages and paragraphs in which they appeared in *Obsession*:

> *Manipulation, domination, control. Knows how to get inside victim's head. Background of virtually all of them came from abusive or otherwise several dysfunctional background, but that doesn't excuse what they do. The sadistic killer anticipates his crime. In fact, he has perfected his MO over his criminal career. Signature aspect—better than MO. MO is what an offender has to do to accomplish a crime. The signature, on the other hand, offender has to do to fulfill himself emotionally. Voyeurism, which would be consistent with hunting, getting ready for next assault. May take photographs or record the scenes. Takes souvenirs—jewelry, underwear.*

And so on and so on. It was as if he was compiling a checklist to make sure he was fulfilling all of the important descriptors of a serial killer.

On subsequent pages Rader had listed the names of other killers, including Ted Bundy, Son of Sam, Ed Kemper, Steven Pennell (a sadistic predator from Delaware who'd already been executed by the time *Obsession* was published), the Buffalo Bill character from *The Silence of the Lambs,* and Gary Heidnik, upon whom Buffalo Bill was partially based. Rader also included a column for BTK. Down the left side of each page he listed the serial killer traits and put in a yes or no as to whether each individual possessed that specific quality.

In his own case, for "Overbearing Mother" he had written "1/2."

Corresponding to "Arrogance," "Self-Centered," and "Inside voices," he had put "No." For "Intelligent," he'd awarded himself a "Yes." What I got from this unusual document was not truth, as I was hoping for from Todd Kohlhepp, but an accurate portrait of how a hideous and vicious serial killer saw himself.

And ours weren't the only books Rader had read, I learned. He had become a "student" of serial murder. He read a wide variety of true crime books and highlighted sections that basically fit his profile.

Did this make him a "better" or more effective killer? No. This question comes up all the time. You don't get better at killing by reading books like ours. But you may gain some insight about a killer's mindset and mental makeup, and that was one of the things Rader was clearly after.

What is interesting about Kohlhepp's responses is that the information he provides, the detail he gives, and his tone are consistent with what he said to detectives during his several interrogation sessions as well as in his conversations and written communications with Maria. Unlike so many other violent criminals I have studied, he is not trying to present a different persona, depending on his audience or what he has to gain with each listener.

The first few sections of the protocol document deal with background information such as date of birth, height, weight, physique, race and ethnic origin, appearance, marital status and history, education, military record, employment record, and medical history, including psychiatric history and any suicide attempts. Chronic behaviors and sexual behavior history cover everything from family structure and environment to any physical, emotional, and/or sexual abuse or trauma the subject suffered, to his own childhood behaviors such as nightmares, running away from home, chronic lying, destruction of property, abuse of alcohol or drugs, and the so-called homicidal triad of chronic enuresis (bedwetting), fire-setting, and cruelty to animals or other children.

One interesting element that arose in this section was his admission that when he was arrested, he was making about $350,000 per year—extremely rare for a serial predator—but that he managed to zero out adjusted gross income on his income tax returns so that he paid only a few thousand dollars in taxes. What this tells us is that people who violate the law in one area tend to violate it in others as well.

It was when he got to the section entitled "Offense Data" that Kohlhepp noted he needed additional pages to answer the questions completely beyond the space allotted in the protocol document.

As I had originally suspected, he said he had told someone else about the Superbike murders. This piece of information is extremely important from an investigative standpoint. If we believe that an UNSUB has talked about the crime with someone, we can go public with that information and sometimes induce that person to come forward, if for no other reason than that person would now be in danger. In this case, Dustan Lawson was a former boyfriend of Kala and the one who introduced her to Kohlhepp. Kohlhepp said that Lawson did "odd jobs on a variety of situations" for him, the most important of which was providing weapons, since with his criminal record, Kohlhepp couldn't legally obtain them on his own. Kohlhepp wrote that Lawson "knew of bike shop," "knew afterward" about the Coxie murders, "helped obtain pills for Kala, knew of her being held by the next day and was paid to release her dog from apartment, lied to me and didn't do it." Lawson denies having been told.

Kohlhepp also said he told his longtime mistress. She is on record as saying she wonders if he thinks he told her, but if he did, it was in some sort of code that she didn't understand.

Note that this betrayal of not letting Kala's dog out is effectively given the same weight as the murders. Kohlhepp never gets over being betrayed on any level. More significantly, anyone who knew about the Superbike murders and had acted on that knowledge as we had hoped he or she would could have prevented at least three subsequent murders.

I was also not surprised to read in Kohlhepp's account that Lawson was not the only one he approached, that he felt a greater need to talk about it. Unfortunately, it didn't quite work out: "I did try to partially confide in a friend who is family and church-minded many years later, 2012–15, to ask help to get my life right, but I beat around topic so bad he didn't know what I was talking about."

As for Kala Brown's disappearance, Kohlhepp wrote: "Dustan was immediately suspicious. People around me knew something was off, but not what."

Other proactive strategies would not have worked with someone like Kohlhepp. While he conceded "I watched news/newspapers online multiple times each day" about the crimes, as I knew the UNSUB would, "I kept zero souvenirs other than a rifle I brought back from a Juarez trip. I did not communicate with any families, police or media. I did not interject myself into investigation."

In the protocol document, there is a long column of descriptive words associated with the subject's crimes, and columns under which to assign numbers corresponding to how significant a factor each of those descriptors was for each crime. The first page of the section has columns for the rape when Kohlhepp was fifteen, the Superbike murders when he was thirty-two, and the murder of Charlie Carver and the imprisonment of Kala Brown when he was forty-five. For example, the first set of words is *angry, hostile*. Under each column, he wrote a 1, meaning anger and hostility were "predominant" in each instance. On the other hand, for *desperate* and *lonely*, he wrote 5's, which correspond to "not at all/absent."

The most interesting variation among the three crimes corresponds to the words *calm, relaxed*. By the time of the Carver murder and Brown kidnapping, he says that was his predominant mood. It was a 2 (significant) in the Superbike murders. But going back to his first serious offense, the kidnapping and rape of the neighbor girl, he rated it a 5: "not at all" relaxed, showing he became considerably more comfortable with violent crime as he got older.

Unlike most predatory killers, his *excited* level was 4—minimal—when he committed a crime, and as far as being *frightened, scared, terrified*, that was somewhere between minimal and not at all. He had minimal sorrow for killing the bike shop victims or Carver, but it was a predominant emotion in the initial rape that sent him to prison. He was also significantly *depressed, unhappy, sad, blue* for having raped the neighbor, but only minimally so for what he had done to Charlie and Kala and not at all for the bike shop murders.

Maria states that the most jarring thing he ever said to her, which she will never forget, was, "You've got to understand, for me it's like washing the car or taking the trash out."

One question on the protocol form asks, "What was the nature of conversation during each offense?" It was intended to bring out both strategy—was there a ruse, con, or enticing conversation used to get to the victim or was it a silent, surprise, blitz-style attack—as well as signature aspects—was there a "script" the offender liked to follow? Kohlhepp responded: "When killing I am silent and focused on what I am doing. Any comments are short, clear, to explain what I need, and calm."

He proves this when he explains: "Bike shop conversation was to help place targets so I wouldn't have to take on all four at once. I went in with confidence I could handle 4–6 [individuals], but not if bunched up if they were armed, so I manipulated a controlled path."

While he claims not to be interested in sexual control over others, maintaining overall control of any situation is clearly vital to him:

> *There was talk only to Meagan after I shot Johnny when she started to panic when that [the alleged intended robbery] didn't go the way they planned, her pleading I not rape or hurt her. I calmly informed her I would not, but I needed to secure her and search her for drugs/weapons. Panties stayed on, conversation was polite and inquisitive, asking her about herself and Johnny. No threats, no insults.*

When it came to describing "evidence of precipitating stress or crisis," which included financial difficulties, family problems, injuries and illness, employment issues, or death of a friend or relative, Kohlhepp assigned nearly all 3's: "moderate/somewhat." With most repeat killers or predators, these stressors are significant precipitators. In Kohlhepp's perception, the only times in which the stressor was predominant were "conflict with parents" in the neighbor rape and the Superbike murders, and "conflict with a female significant other" with the rape.

Altogether, Kohlhepp's numerical responses showed a straightforward acknowledgment of what he had done and no attempt to mitigate the wrongness of his actions or to blame others.

For example, he described flying down to Juárez, Mexico, to hunt drug dealers with a group of other hunters as being

> *like a very bad movie. Nothing cool about it. All type A's who paid a lot of money for top of the line hardware and tactical training, dreams of being a SEAL team. We were anything but. Mostly a bunch of heavily armed jackasses that wanted to try out their toys and kill something, targeting drug dealers was just ethically acceptable to group.*

Whenever another person figured in Kohlhepp's answers, such as his parents, his take on events struck me as realistic and accurate. He may have been in the dark about *why* he did certain things, but the *what* and *how* rang true.

This is not to say that he saw himself as a predator. In his mind, he had developed a relationship with certain of his victims beforehand. Under a question about what sexual acts were committed during the crime, and in what order, Kohlhepp felt a need to write in that with the neighbor girl, they had engaged in "petting/light fondling once or twice before any of this happened." [The victim denied this ever happened.] For his most recent crime, he wrote, "Kala was met at strip club, became my prostitute. [She denies this.] Sequence started with dinner or her telling me what bills she needed paid."

Under "How does the subject continue to maintain control of the victim for repeated assault?" he added:

> *Kala needs to be explained. Presence of weapon and I showed her the grave I dug for her when she acted out over drugs. She was not very scared when I shot Charlie, more confused, then quickly turned it into what could she get out of it. She was turned on and explained some [fetish]/submissive fantasy of hers. I just didn't want her running to cops. I think she would have ran to her dealer first. Chains and container overnight were for my peace of mind, she was loose during most of day but I did keep weapon when she wasn't chained. As long as I kept buying her stuff, giving her attention and pills for her to crush and snort, she seemed content. Charlie was rare issue for her, only what I would or wouldn't buy for her. She wanted sex, attention and [drugs]. The [drugs] I refused. When I rejected sex a few times, it pissed her off and she was not happy I wouldn't play along with her submissive fantasy. It's my fault for all this and it was wrong. I am only saying she pushed the fantasy part, not me.*

LIKE MANY RESPONSES WE GET FROM VIOLENT OFFENDERS, THIS ONE NEEDS SOME nuanced interpretation to be meaningful and useful. First, let us stipulate that no matter what Kohlhepp says about Kala's reaction to his shooting Charlie, she had to be terrified and fearful for her own life, particularly after he showed her the grave he had dug for her. Second, whatever she asked for or however she behaved was unquestionably some form of survival and coping strategy. Without even getting into the question of whether she requested drugs from him, if they had had a sexual history with each other prior to his taking her prisoner, it would be natural that she would revert to whatever roles they had each assumed, since she would be attempting to "normalize" their relationship so he wouldn't see her as a threat that had to be eliminated, as Meagan Coxie had been. And for the length of time she was held before her rescue, I was not surprised to hear that she was "pissed off." No matter how fearful she might have been, you can't be a captive for that long under such uncertain circumstances without real emotion coming to the surface. *Content* is not a term I would see her using to describe herself during the weeks she was held in the Conex container.

That all being the case, what is Kohlhepp actually telling us here?

First of all, he is asserting that since his first ill-advised sexual assault when he was fifteen, he does not consider himself a rapist or a sadist. Contrast this to Joseph Kondro, who had no problem with the term, since it identified him precisely.

Even though they were both killers, Kohlhepp would have nothing but contempt for the Kondros of the criminal world. He doesn't even consider himself kinky in any way. "She wanted me to role play on various levels and I would not." In fact, in response to a later question he stresses, "Chains/handcuffs were control only. No foreign objects, whipping or spanking," despite his perception that Kala had domination fantasies in which she wanted him to participate. He admits that he "did place a trail camera in container for the first week to see what she was doing," but insists it was "to review security, not voyeurism."

His self-image is extremely important to him. He readily admits to murder and is even "man enough" to say it was wrong, but it is not part of his sense of himself that he would force himself on a woman. He wrote: "When Kala got ugly, I just left her alone and did work in another area." In other words, the only punishment he imposed on her was withholding his presence. And in answer to a question about "Evidence of sexual dysfunction during the assault, he writes: "No sexual dysfunction."

Even admitting to sexual assault on Kala has to be explained away as expedience: "Admits fully on murder and kidnapping. Not guilty on rape with Kala, but not worth staying in jail [before trial and sentencing to prison] all that time to fight a charge that has no impact on my life. What was the point?"

To further make his point, in a later question about whether his offenses against Kala were sexual or nonsexual, he points out that "rape victims do not request vibrators and stripper poles," which he said Kala had asked for and he procured for her during her imprisonment.

It is as if he is saying, "Yeah, I imprisoned an attractive young woman, killed her boyfriend, and kept her chained in a container in the woods where I had repeated sex with her, but it's not because I'm a pervert; there were practical reasons for all of this. Oh, and as far as that sex, it was mutual, and when either one of us didn't want it, we didn't do it. Not only that, I bought her whatever she wanted." Just as in much of the interrogation, he is showing how "reasonable" he is.

In response to "Sadistic acts committed during the assault?" where a simple 3 for "absent" would have sufficed, he wrote: "No sadistic acts. Not that any appreciated it, but methods and ammo was carefully chosen for efficiency so [as] to kill rapidly and while that was primarily to protect me, it did reduce their suffering. Enhancing pain is uncalled for."

Later he writes: "Violence does not turn me on, neither does control."

There is a place on the form for "any other big changes?" meaning stressors that could have influenced the decision to commit a violent crime. For this Kohlhepp needed considerably more space than was provided: "Constant issues with being on sex offenders website with company work, under heavy stress constantly. Grandmother died day I got bike—mom/grandfather fighting." And for the first rape: "Issues with Dad and physical/verbal abuse constantly."

The family issues are definitely stressors, but I think the sex offender label is one of the keys to his adult personality. No matter what else he accomplished, Kohlhepp perceived that it was a blot on his life that had ruined everything. And he wasn't completely wrong about that.

Though the rape in Tempe was by far the oldest crime, it is the one for which the most emotion and passion comes through. It is the only crime for which he expresses genuine remorse: "Wish I could say otherwise, but I really don't feel bad about most of the others." He says that he had been drinking for the first time, having sex for only the second time, and like most sexual predators, he found that "it felt good, but whole situation did not and I knew everything was wrong, was not a turn-on." He would be "surprised if I lasted a minute." The fantasy is almost always better than the act for these guys.

Even though he writes that he doesn't feel bad about the other crimes, we can clearly see that there are elements of conscience warring within Kohlhepp, as opposed to someone like Kondro, who has absolutely no qualms about what he did, or Harvey, who has to justify each murder. Only a few lines later he writes: "I hate that I killed the mom at bike shop and wish I had avoided her. Johnny is good riddance, brought it on himself. I really wish I had come up with better, nonviolent answer for Meagan. I am sorry for Charlie and Kala, I overreacted and should have just fired them both. I didn't care about Kala so much, I just hoped for a different outcome than Meagan, but was probably going that direction."

As conflicted as he is about much of this, there is real introspection here, rather than an attempt to explain it all away as someone else's fault. Late in the protocol he wrote: "I was on a bad path and determined to continue. I can sit here saying this or that could have prevented this, but it all comes back to me. I should have stopped me."

Dealing with a section headed "Assault History," he answers the question "Evidence of increasing use of force or aggression over time in his assaults?" with "Yes—murder became common."

Other than that, he says he doesn't see any similarities in his victims corresponding to a list of traits ranging all the way from age to race to hair color to physical handicap, or even motivations for his crimes.

Then he adds a comment that doesn't correspond to any specific question on the protocol: "Looking over my past, I don't see a pattern, it was the Swiss army knife of killing, whatever tool fit the situation. I don't see a so-called signature or constant, which confuses me, and I am sure, others."

This is a pretty accurate self-appraisal. Yet he knew something was wrong. "What would subject have liked friends, relatives, or workmates to do once they found out?"

He responded: "Help me find therapy, especially at first [the rape in Arizona]. I was asking for help at the end. I wanted to get back on the right path. Once Meagan/Johnny [murders] I really was back on a wrong one and without something/someone pausing it, I don't think it was going to end without help by courts or friends stepping in."

25

ORGANIZED VERSUS DISORGANIZED

Going all the way back to the rape of his teenage neighbor, Kohlhepp displayed both organized and disorganized elements in his crimes. No matter how careful their planning, for most violent and predatory criminals, since crime is essentially an irrational act in any society that prohibits it, there is often a point in which logic and reason breaks down.

In 1981, John Hinckley Jr. had a plan to impress actress Jodie Foster with his love for her by assassinating President Ronald Reagan. Since it was a crime against the president of the United States, the FBI took primary jurisdiction and my unit got involved in preparing the case against Hinckley. The point of all this is that the shooting itself, though thank God no one was actually killed, was well planned and executed. The next part, not so much. Foster was supposed to be so impressed that she would be drawn to him. Then he was going to demand an airplane be provided to fly the new couple away. That's where his planning and logic ended.

And we can see the same traits in Todd Kohlhepp, even though he seems to me a lot smarter and more practical than Hinckley, who was found not guilty by reason of insanity in a verdict that remains controversial to this day. Yet because Kohlhepp is different from Hinckley in personality and intelligence, I saw other significant differences that I was hoping our questions would elucidate.

With Kohlhepp, the balance between organized and disorganized seemed much more organic than haphazard. With him, I began to realize that whenever he went too far in one direction or the other, something inherent in his psyche would steer him the other way. This emotional push-pull, in which he got himself involved with something and then questioned himself about it, was a significant behavioral indicator. And in that, he was completely unlike true predators, who don't question either their motives or the way they go about their crimes.

We see this in Kohlhepp's ventures into Mexico to kill Juárez drug dealers. For a while the idea made sense to him, but after a few trips, he realized this action was stupid, bordering on crazy.

The element of Kohlhepp's record that interests me most from the standpoint of a breakdown in criminal logic and planning is the abductions of Meagan Coxie and Kala Brown. While he may have felt in retrospect that the shooting of Johnny Coxie was justified and that of Charlie Carver was an overreaction, the fact remains that in each case, he cold-bloodedly and impulsively murdered someone and was left with an eyewitness who could bring him down.

"I met Meagan on the corner of Blackstock and Reidville Road by bridge over I-26, panhandling," he explained to Maria. "Cute girl begging for money. I offered her a job cleaning homes and honestly thought I would probably get some sex out of it for helping her. Didn't meet Johnny until the morning of shooting, but she had informed me of him as boyfriend, not husband."

With his strange self-defined and governed sense of morality, he didn't want to kill Meagan or, later, Kala, which would have been the most efficient way to eliminate the witnesses. But what to do with them?

He could hold Meagan for a little while, but then realized he would have to do something. Other than killing her, was there a means of getting her out of the way so she wouldn't turn him in and testify against him? After talking with her extensively and learning as much as he could about her, Kohlhepp came up with a plan. And in its own way, it was almost as farfetched as John Hinckley's.

"The Conex was not meant to be a cage," Kohlhepp told his interrogators. But once Johnny Coxie pulled a knife on him with the intent to rob him, he said, "I shot him." Then, "I didn't know what to do with her. I didn't want her in my Conex because I had stuff in there and didn't know what the hell to do with that. Putting her in with my guns is not a good move. For the first time, I was having a little panic of what the hell to do with her—put her here, drop her, what the hell do I do?"

The repetition, I think, shows how uncomfortable he felt being out of control.

I do believe he did not want to kill Meagan, but he was facing an insoluble problem. Johnny's killing "really bothered me, 'cause it was such needless bullshit. Um, hell, I was getting them money; why were you robbing me?"

With some effort, he says, he calmed her down, handcuffed her, and left her on the floor of the Conex container. After he dug a hole and buried Johnny, he went back to the container with food for Meagan.

"You fed her after she tried to rob you?" one of the detectives asked.

"Well, what are you gonna do with her?" Todd replied. "I didn't want to shoot her."

He went on to explain, "She was talking to me and first she had drug issues, and then she kept going off the deep end with weird shit and kept talking, and then she kept telling me that she had manic . . . manic modes or sort of bipolar lithium crap; I don't know what the hell it was. Lord, she was up, down, up, down, up, down."

This continued for "five or six" days, during which time Meagan proved to be an unreliable prisoner. Kohlhepp reported that he bought her cigarettes and she tried to burn out the metal container. "I come in and often find that she burned this. She's sitting next to 100,000 rounds of ammo. Love of God, please stop burning shit!"

But during this time, he came up with a plan. With Meagan's perceived drug problem and what he said was her trouble with the law ("I guess you guys had arrested her for meth or some shit," he commented to detectives), combined with his need to have her disappear, Kohlhepp offered her a proposal.

"[I] took her out the end of the building, sat her down for a while, got her to calm down. 'Just please calm down!' Got her some food, told her basically that if she would just chill the hell out . . . 'You don't know me. You don't know very much about me. You don't have shit.' And last time I could check from what was online, she had a warrant already looking for her ass. Uh, 'I'll give you $4,000. I'll drive you up to damn Tennessee, I'll drop your ass off somewhere. If you've got any common sense on this planet, you'll go left and I'll go right.'

"I told her I would give her $4,000 and basically release her in Tennessee: 'Just go, please go, don't come back.' It seemed an easy . . . so it seemed like an easy solution. She didn't know my name, she didn't know my address, she didn't know where I lived."

We see a similar duality in Garrett Trapnell's decision to hijack the TWA flight in 1972, a crime it is virtually impossible to get away with, especially when his trump card was the demand for a pardon from President Nixon, a completely unreasonable expectation. Yet the "Free Angela Davis" demand to ingratiate himself with his prospective prison mates shows a high degree of sophistication and planning.

So what happened to alter Kohlhepp's "easy solution"? He spun two narratives at this point that were not entirely in concert with each other, not surprising since he was basically improvising as he went along. I think this also demonstrates that logical push-pull going on inside his head. During the interrogation at the detention center he told detectives:

The weather went to shit. We were having sleet, it was right before Christmas, man. We were having sleet, we were having rain, the weather went to shit, and I still had to find a way to get away from Ashley, my girlfriend. I would have to get up out of work, get this person to Tennessee, drop her off, and get home. That's not a . . . that's not just a couple-of-hour trip. And I'll drop her, ain't gonna be at the border. We're going north to Nashville. I want her way away from me. I want her to forget where South Carolina is . . . "If you've any common sense, you'll keep walking, go get a job at a, at a diner somewhere, work as a waitress, get your shit right, don't come back. They're gonna watch you for a year, and I'm here, so don't come back." She was gonna take it. She was happy, she was happy as hell for, like, two days. I just couldn't get past the weather.

BUT A LITTLE LATER, HE ALSO TOLD THE DETECTIVES THAT MEAGAN SET ANOTHER fire, and that was why he decided to kill her:

When I went into the building, I mean, I was choking. I went, "Holy shit!" I went to get her out, and then all of a sudden, it's like I had a caged animal. Oh my, I don't know what the hell, the hell, she went from "I'm so friggin' happy in the world to be, I'm gonna go to Tennessee with money and I'm gonna restart my life and thank you, thank you," to batshit crazy.

At that point I tried to walk her out of the building. I just had enough. I walked outside, I was trying to calm her down, figure out what the hell to do with her, what to do, what to do with her. I didn't know. Um, back, came back in the building, um, she was going nuts, just . . . It wasn't like she was emotional about the situation. This, this had been days. It wasn't just about that. It was just like serious chemical imbalance shit. And she walked outside. I walked, I walked her outside. I walked her outside, I put a bullet in the back of her head.

BOTH OF THESE EXPLANATIONS PROBABLY FACTORED IN KOHLHEPP'S DECISION TO KILL Meagan Coxie. But the psycholinguistic analysis of her part of his narrative—the obscenities, the repeat words and phrases, the frequent acknowledgment that he didn't know what to do next—points directly to the reality that knowing he could never be in control of the situation and that in her own way she had taken advantage of him was what ultimately led him to another murder.

This was Todd Kohlhepp's behavioral pattern. It wasn't that he was punishing Meagan for trying to rob him; he had already made his point by shooting Johnny. It was that she had put him in an untenable position, and he didn't know any other way to react. This was often when the organized-disorganized behavioral pattern manifested itself.

Each time he killed, it was because he felt he had been put in such a position—whether it was the two men in the Hunt Club apartment parking lot he said he killed after they had tried to rob him, the people at Superbike who had ridiculed him and, he believed, made fun of him, and then Johnny and Meagan Coxie. And there is little doubt in my mind that the same fate would have awaited Kala Brown had she not been rescued. He would have known it was wrong, but the logical, organized part of his brain wouldn't have seen any other "sensible" way out.

I think someone as intelligent and analytical in his own way as Kohlhepp would have come to realize that his plan for releasing Meagan in Tennessee was far from foolproof. If, as he suspected, law enforcement was looking for her and eventually found her, she would have used any information she had to trade for more favorable treatment. And even if that part was not true, most violent criminals have learned that the only way to ensure that a secret remains a secret is if only one person is privy to it.

Because ultimately, had Kohlhepp considered how to deal with Kala, he would have been faced with the same haunting question that confronted him with Meagan: *Now what?* His response on the assessment protocol makes this abundantly clear. In answer to "What did subject think about after offense," he replied, "Here we go again. Holding Meagan was huge mistake, why the hell am I holding Kala? Drop her and clean scene, remove evidence."

That we see a mixed presentation in Kohlhepp—shooting Charlie and dealing efficiently with his body and car but not knowing what to do with Kala—is also not surprising in someone who has become criminally sophisticated, but whose main focus is on his business and everyday life, as opposed to someone like Dennis Rader, whose BTK crimes were the center of his existence. Rader's key motivation in life was perverse sexual gratification achieved through murder. Kohlhepp's crimes had to do with irrepressible anger and retaliation.

Compare how Kohlhepp reacted in this situation with Rader's response. Rader took immense pleasure and satisfaction from his "projects," as he called them. He could dominate his victims and decide their fate, and he had the mystical feeling that by killing them, he would possess them all as sex slaves in the afterlife. Rader's greatest regret was that he couldn't spend *more* time living out his sadistic fantasies with his victims.

Kohlhepp, on the other hand, was at his wit's end when he had to deal long-term with Kala. His account and Kala's are understandably different. She describes weeks of horror and uncertainty during which she daily feared for her life, while he describes a needy and demanding prisoner who would not leave him alone. But what is clear in the assessment protocol is that this is no Gary Heidnik, keeping women captive for his sexual gratification and kingly fantasies. Unlike Heidnik or Dennis Rader, whose sadistic crimes against women were the center of their beings, Todd Kohlhepp focused on his real estate business. In Heidnik's case, the captivity is organized. In Kohlhepp's, it is disorganized. Yes, he liked sex and pornography, but he had far easier and less fraught ways of satisfying himself. This is a man who has gotten himself into a situation that he doesn't know how to get out of, as he revealingly illustrates in the protocol:

> *She started in with this submissive kitten nonsense, wanting TV, books, then blue hair dye and vibrator, pills constantly, became bored and wanted sex, and me to entertain her, using sex to trade for things she wanted. Watched TV, complained about no* [*drugs*] *and why I didn't show her more attention, only a few comments about Charlie or anyone else, just herself. She did ask me to kidnap a girlfriend for her to have like a pet and I refused. Having a captive is not fun at all.*

Compare this confession to his self-assured narrative to the sheriff's detectives on how he went about the Superbike killings, where, while he admitted what he did was wrong, he took a measure of pride in the organization and efficiency of his actions. When it came to holding women against their will, he was in way over his head and learned nothing from his own experience.

And in implicit rebuttal to the Heidniks and Raders of the criminal world he wrote, "I do wish I could have found a better way to deal with Meagan, but having a captive is extremely stressful and I have no idea how anyone could be turned on by that logistical nightmare, emotional mess."

26

NATURE AND NURTURE

Todd Kohlhepp's life story and the way he answered questions fell in naturally with how the assessment protocol was structured: Background Information, Chronic Behavior Patterns, Family Structure and Environment, Assault History, Juvenile Record, Victimization History, and Subjective Assessment of Behavior.

Although like Joseph McGowan, Kohlhepp had had major problems with his parents, Kohlhepp didn't lash out at an innocent child in a moment of displaced rage. Both Joseph Kondro and Kohlhepp were irritable and angry children, but Kohlhepp didn't wantonly rape and murder those close to him. And although like Donald Harvey, Kohlhepp preferred being in total control of his environment, he didn't arbitrarily kill easy targets just because he could.

Kohlhepp was a different sort. No less a product of his circumstances, but a killer who could have had a different outcome. With my *This Is Your Life* approach, here is how I see it. The quotes in this chapter are taken from the interrogations, the communications with Maria, and the assessment protocol, all of which, as we've noted, are remarkably consistent. Other than his bragging about his proficiency with guns, none of his statements are self-serving, nor does he try to place the blame elsewhere. Kohlhepp is a *believable* killer.

Todd's parents divorced when he was less than two years old. In fact, he says, "Dad was on a date [the] night I was born. [His father denied this.] He spent his life chasing women, big dreams of success, very aggressive and violent. Very intelligent, mean as hell." His mom, Regina, known as Reggie, married Carl Kohlhepp the following year.

Todd reported that he got along well with his two stepsiblings—Michelle, who was a year older, and Michael, a year younger. "We had moved to St. Louis for Carl's employment," he told Maria, "and while we were in school, the kids got kidnapped by their mom. Their mom came up, called the school, informed them that she was the aunt; that the mother had been killed, meaning my mother had been killed. That she was the aunt, but that she was only there to pick up those two kids, not me. The school never bothered to even question it." Todd was seven at the time.

"She just took them," he wrote, "left me, informed them that my mother had died but that she only wanted those two, and then drove back to Georgia. Hours later I got picked up. Nobody could figure out where the other two kids were. When it finally came out, they realized that the mother had taken the kids and Carl did not want to have her prosecuted, so he went ahead and relinquished custody, which entirely changed the dynamics of the household, because from that point on, Carl's attitude was hostile, on a good day."

After Michelle and Michael were gone, "I spent my time in my room. Nobody wanted to talk to me. Nobody would tell me anything."

A probation report from the time of his arrest for rape at age fifteen stated that "he has been unabatedly aggressive to others and destructive of property since nursery school." He was nine when he spent time in the Georgia Mental Health Institute for being aggressive toward other children.

Kohlhepp doesn't deny his preadolescent behavior. "I spent a lot of time with the school counselor; had a lot of problems at school. I'd act out a lot. With me constantly moving from school to school to school, I never really had friends. Then, because I was always the new kid, I would really find issues with bullies. Then, I would take abuse and then basically it would build up and then I would act out. Typically, when I would act out, I would hurt several people. I usually took it to an extreme: *They won't do that again!* But because of that, I spent a lot of time with the counselors." And what did these counselors do for him? Was the stint in the mental health facility productive, and was there any follow-up therapy? There is no evidence.

When asked about his happiest memories, he replied, "Playing with animals on grandparents' farm in Georgia. The animals I liked, grandparents I didn't. At age five my grandfather thought it was fun to hit me with cattle prod. Not a fun experience. At seven he would castrate pigs and threaten me to be next. Eight to nine I was dragged to tree, tied to it, and beat. I'm surprised I didn't kill him. He had no friends and took pleasure in hurting and controlling others. He treated his daughter (my mom) same way."

He said that for a while he was shuttled back and forth between his mother's house and his grandparents. "It was a very weird situation. They had never wanted to have kids. They made it clear they never wanted to have kids. They reminded my mother constantly that she was never wanted, and of course, that filtered down to me."

The life he described was bleak. "On and off, my mom would take me for a while. Then she'd take me back [to my grandparents]. While I was living there, if my grades weren't good, if I had done anything he didn't like, if I hadn't woken up and cleaned the chicken coop at five in the morning the way he liked it, it didn't matter what it was. I would get woken up, dragged by my hair out back, tied to a tree, and then beaten. Usually it was with a large leather belt. If he didn't use that, one time he actually used a horse crop."

Since I've been doing this kind of work, I've had to learn to compartmentalize the various aspects of the lives of the killers I've studied. By that I mean, I familiarize myself with every detail of their crimes and loathe what they did. At the same time, I may feel tremendous empathy and sorrow for what they went through in their young lives that contributed to their adult behavior. No kids should be treated the way Ed Kemper and Todd Kohlhepp were. And it is easy to see how Kohlhepp's family background—growing up with little love or affection and having his stepbrother and -sister suddenly yanked out of his life while he was overtly rejected—would have contributed to his inability to develop a trusting relationship with another human being.

When his mom, Reggie, wasn't with Carl, Todd claims she was always looking for another man in her life. "I guess she thought she would get a man quicker if [there] wasn't a son running around."

By the time Todd was twelve, he'd had enough of living with his mother and grandparents and announced that he wanted to go live in Arizona with his biological father, Bill Sampsell, whom he had not seen in eight years and hardly knew. Around that time, Reggie bought him new bedroom furniture, which she thought would make him like his home environment better. Instead, he quickly destroyed it with a hammer. He commented that he thought it was "girlie."

"A little drastic," he conceded to Maria, "but I was twelve. I acted out a lot at this time. No friends, I wanted far away from family—seemed like a good idea at the time. Mom was frustrated with me and my not accepting whoever she was dating."

Finally she agreed to send him out to Sampsell.

"So I went from [the] conservative Bible Belt South to Tempe, Arizona, which is hot, wild, and filled with college coeds. Dad owned a restaurant—Billy's, Famous for Ribs, Tony Roma knock-off. He was busy with that; I had almost no supervision. He started fights with everyone, chased every female with a heartbeat, and the little time I was around him was not what I expected. Man would go from zero to violence like a light switch. I stopped being so shy, violence level increased."

The way Bill Sampsell described it to Maria, Reggie "just shipped him out there and gave notice after the fact that she had already put him on a plane."

Todd became enamored of his father's weapons and listened to his stories of having served in the Special Forces, which turned out not to be true.

Unsurprisingly, Todd's and Bill's recollections of the time the son spent in Arizona are quite different. "I thought at one point we were, we were at, you know, fairly decent equilibrium. I didn't think there [were] any problems," Sampsell told Maria.

Then Todd raped his neighbor. The key to Kohlhepp's evolution as a killer is this rape when he was fifteen. In terms of precipitating stressors, he related on the protocol:

> *Dad was gone for week out of state, due to return that night. I knew I had a beating coming for something upon his return, messing around with liquor cabinet and counting down the hours. I was upset and frustrated, really only intended to talk with [the girl], convince her I was the right one for her and totally screwed that up. I don't know why it escalated like that, but it did. I wanted someone to want me.*

I would make the case that if it hadn't been for this one incident, Kohlhepp's life would have been completely different. That doesn't mean I condone what he did, because of course I don't. With other offenders I've studied, including the three others in this book, I see a certain inevitability in their development, even for someone like McGowan, who was stopped after his first offense. But let's go through the case history to support the point about Kohlhepp's life of crime.

First, on the protocol question of whether anything would have stopped him from the assault, he answered:

> *Neighbors, teachers, school counselors and employees at my dad's restaurant knew what all was going on, what I*

> *was dealing with and that I was in trouble and becoming [a problem] quickly. None of this happened behind closed doors. Someone could have stopped at any time, taken an interest, got me some help. Instead, they passed it on to the next person. I was open to therapy at that time.*

This is not an uncommon observation from a convicted felon, but in Kohlhepp's case, I think it has particular significance, because his other responses from all three sources are so candid and straightforward that I believe he is not trying to avoid responsibility for any of his crimes. He is merely perceptive enough to recognize where the inflection points in his life were that could have made a difference.

To demonstrate how confused and disoriented he was after the assault, his victim says he contemplated killing her to keep her from talking, as he later killed Meagan Coxie. He writes, "I was mostly polite and apologized to her, but I did threaten her family," he conceded. Again, we see a mixed behavioral presentation. He knows what he did was wrong and feels almost immediate remorse. At the same time, he doesn't want to face the consequences of his behavior.

In response to a protocol question about how he behaved after the offense, he wrote: "In shock. Not sure what just happened, not sure what was going to happen. I released her, my dog got out, and she helped me get the dog until cops showed up. She went to them, I went home.

"When cops pulled up and found her," he recalled, "I left and went to my house, refusing to come out until Mrs. Taylor two doors down talked to me and I put down the gun. I couldn't wait for cops to take me; I did not want to be near Dad. I mostly felt disconnected and scared, remorse came later. I just wanted out of that house and away from all family. I was scared of my dad, embarrassed over what I did, and numb, not knowing what to do, where I was going."

He couldn't even admit to himself that he had committed a sex crime despite what the girl reported, and he continued to stick to the story about the lost dog. Responding to a question about what he thought after the offense, Kohlhepp wrote: "Embarrassed, avoided topic at all cost." As to whether he tried to "avoid being detected by law enforcement," he replied: "Didn't even try. I was grateful officers removed me from house. [I] had no clue I would get fifteen years, thought I would go to juvie until eighteen and counseling."

My sympathy and empathy are always with the victim, not the offender. At the same time, it should be noted that in cases of sexual assault where the victim is a teenager, one of the considerations in sentencing, aside from the degree of violence of the attack, is the age of the assailant relative to that of the victim and whether they knew each other. If the ages are close, there is often a show of leniency, of the possibility of a second chance, and an attempt to determine if the crime is part of a predatory pattern or an isolated incident from which the perpetrator can be turned around with the right attention. I don't think such an approach would have been inappropriate in this situation.

Instead, Kohlhepp was tried as an adult. "Approximately six years of intervention in fifteen years of life have resulted in abysmal failure," wrote juvenile court judge C. Kimball Rose of the Maricopa County Superior Court in transferring the case to the adult division.

The plea deal Todd agreed to got the rape charge dropped, but it sent him to prison for fifteen years, the amount of time he'd been alive. And it left him with a sex offender label for the rest of his life. It is as if once he had committed this single act, his entire life was unalterably tainted. And this is a genuine shame and loss—of at least seven lives, among other things—because if Todd Kohlhepp had been handled properly, the outcome of the situation might not have been preordained.

There is a common and politically fashionable belief in some quarters that "rape is rape," that every sexual assault is alike. And while every sexual assault is unquestionably horrible, they are not all alike and the men who commit them fall along a continuum.

In our research at the FBI, we identified several basic rapist typologies and the crime of rape into more than fifty subgroups. Though in the latest edition of the *Crime Classification Manual* we break rapists into about a dozen categories for the benefit of investigators, for our purposes here, there are four typologies that cover most rape situations: the power-reassurance rapist; the exploitative rapist; the anger rapist; and the sadistic rapist.

For the anger rapist, also referred to as the anger-retaliatory rapist, whose attacks are a displaced expression of specific or generalized rage, and the sadistic rapist, who gets his satisfaction through the suffering of others, there is little realistic hope or expectation of rehabilitation. The exploitative rapist, who is a more impulsive predator and seizes opportunity that presents itself rather than planning and fantasizing about the crime he wants to commit, is often in the process of committing another crime, such as breaking and entering, when he rapes. He can sometimes be helped if he is caught early enough in his criminal career.

The power-reassurance rapist, as the name implies, tends to be an inadequate type or someone trying to prove his sexual potency to himself. This type may be a loner and even fantasize that his victim is enjoying the experience and might be attracted to him. A large portion of date rapists fall into the power-reassurance category; the remainder are exploitative rapists. One of the key characteristics of the power-reassurance rapist is that because his self-esteem is so low, if you can get to him early enough with therapy, his behavior can be modified.

But that's not what happened with Todd Kohlhepp, whose attack was definitely of the power-reassurance type. Instead of being given therapy or sent to the type of institution that might have tried to deal with his myriad hostilities and antisocial personality issues caused by both his inherent makeup and his unloving and unsupportive family environment, he was sent to an adult prison facility. His prison record shows several years of disruptive and sometimes violent behavior before he started settling down around age twenty.

He claims he was never sexually assaulted or molested by other older prisoners, but this is one of the few times where I'm not sure I believe him. I suspect such treatment would have taken place, at least at the beginning, and would have furthered his distrust of others and his anger at other people.

My focus is on understanding why people commit violent and predatory acts, not to help them become better, more law-abiding citizens, but to aid in catching them, prosecuting them, and putting them away. By the time they come into my orbit, they're usually beyond help. Such is the case with Kohlhepp, but it didn't necessarily have to be that way. Essentially he was a troubled boy who committed his first and only serious crime and was put away until he was thirty in an institution where lack of love, caring, and trust would be reinforced and he would be exposed to hardened criminals of all types. His natural development was essentially put on ice for a period as long as he had been alive when he went in. For that entire time, he could not develop as an individual. He had to be in continual survival mode.

Anyone who knows me knows that I am not at all soft on law-and-order issues, but I firmly believe the case of Todd Kohlhepp was mishandled from the beginning. He had some of the inherent nature to become a criminal, but in his case it was the nurture—or rather, the lack thereof—from his family and from the law enforcement establishment that fulfilled that destiny.

In Maria's interview with Alan Bickart, Kohlhepp's defense attorney in the rape charge, Bickart said he agonized about how to proceed. Todd was "too smart" to be determined to be in need of placement in a mental hospital; he couldn't be tried as a juvenile because the crime was too serious, but adult prison, the only real option, would only make him worse. The system, Bickart lamented, had no good options.

Nor was he given any guidance when he got out: "Releasing an inmate after fifteen years in prison, who went in at fifteen years old, with no parole, supervision, or therapy, no one to talk to if there was a problem, is a huge mistake. Attitudes and solutions coming from prison environment."

When he was released in August 2001, he went back to South Carolina, where his mother was living, because he didn't know where else to go. Remember, this is a thirty-year-old who has had no experience of living on his own in an adult world and has never had an adult job. He has had no dating experience, which explains why he frequented strip clubs and prostitutes. He found a job and started living on his own, and earning a living. He suddenly had to teach himself an entirely new set of survival skills.

Though he continued frequenting prostitutes and strip clubs and relying on online pornography, he eventually developed the social skills to have "normal" relationships with women. He had long-term relationships with two women he categorized as his girlfriend and his mistress but never sought to marry either of them. I believe this was due to his perceiving how disastrous his mother's relationships with men had been and, spending his formative years in prison, he just didn't have the capacity for trust that marriage represents.

In spite of that, he was determined to turn his life around. The protocol asks for ways the subject feels the offense in question "significantly changed him in any way." Upon his release from prison and move to South Carolina, he wrote:

> *I took a hard look at myself now that I didn't have family influence in decisions, worked towards being a better person. No more crime, wouldn't steal or be involved with [anything] shady. Made sure women didn't feel cautious around me and [I] respected the word no. Didn't put myself in bad situations, the whole treat others how you want to be treated. Confidence went up. Took classes to get education with steady As (I was a C student before) very serious work ethic. This was not a ruse.*

I became the person you want dating your daughter or living next door. Once release I was the perfect employee, went to church, respected law enforcement, followed the law and liked by people. Life was good.

ALMOST AS DAMAGING AS THE LONG PRISON TERM BEGUN AT A CRITICAL POINT IN HIS adolescent development was the fact that he was a registered sex offender. I'm not suggesting it was inappropriate; he *was* a sex offender. As someone who has dealt with sexual assault my entire career in law enforcement, I'm in favor of sex offender registries. But it further added to his struggle to establish himself, which in turn engendered more anger and made him feel as if he had to bend the rules and dissemble to accomplish anything in life.

No matter how he tried to improve himself, he couldn't escape the label of sexual predator, except through lies and subterfuge. And even that was particularly difficult, since by the time he got out, there was an Internet, and his name was on sexual predator websites. But as is true with anything else in criminal justice, such websites can be abused by people with ulterior motives. And that is what happened here.

"I was getting a lot of hate mail, a lot of hate phone calls," Kohlhepp told Maria in a phone call. "I had harassment constantly. I mean, it would happen for a while; then it would go away . . . I had Realtors who would call up [my] clients. I had one Realtor that sent out eighty-eight letters to all my clients with a copy of a website informing them that I was on the list—they should check into it and that they should all basically hire her. Although nobody would tell me who it was, they would hire her. I mean, I was constantly getting it. That's the reason why I changed the name of my company to TKA. Originally, it was Todd Kohlhepp and Associates. I was trying to get my name kind of off of it."

"When I was harassed due to sex offender website, I called police and was told I put myself on the list, deal with it. Church suggested I try another church," he wrote in the protocol.

From this point on, despite the success of his business ventures, his apparent outward enjoyment of the fruits of that success, and the high regard of his clients and colleagues, looking back on his life from the perspective of the protocol, he realized he was already on a downward spiral, that whatever was wrong with him was already baked in. The contrast with his assessment of the period after his release from prison is stark. About the period after the Superbike murders, which he had gotten away with, he wrote:

> [I] *was in a very confused state of mind, going to college again, had several girlfriends, a job and thing were slowly moving forward, but on the side I was doing things to wreck it all. Family a mess, stopped with church, obsessed with good grades but had gotten in with arms dealer from another state . . . became very aggressive and overly cautious, paranoid of crowds and blind spots, parking lots, armed at all times and learning new skills. Stopped looking for relationships, focused on getting what I wanted and people were replaceable. Yet several of my affairs lasted 8–11 years, going on at same time and they knew about the others, didn't like it, but they accepted it. I didn't hide it, but also careful not to rub their nose in it either. I focused more on the bad of the world than the good.*

IF YOU CAN GET MOST SERIAL KILLERS AND PREDATORS TO BE HONEST WITH YOU, THEY will admit that they never would have stopped killing on their own. Kohlhepp also admitted as much, but there is no joy or satisfaction in it for him, just a weary inevitability. He related in the protocol that after he killed Carver and imprisoned Brown:

> *I was exhausted with life, burned out with work, going through the motions but not really caring. Avoided girlfriends, friends, employees, basically was becoming a hermit. I went from helping everyone solve all their problems to telling them*

to solve it themselves. "I don't have time for this." Didn't like Kala, didn't want to kill another woman, very stressed about it. Making comments to friends [that] I was happier in prison. I knew how to cover up, beat phone tower ping and zone tracking, I just chose not to bother. Not to be arrogant, but they didn't catch me, I caught me . . . That [sexual predator] website caused me a lot of stress, and between that and the people wanting to steal from me, I am sure another situation would have happened eventually.

OTHER RESPONSES ON THE ASSESSMENT PROTOCOL FURTHER DEMONSTRATE THIS. ONE question toward the end asks, "Looking back on each offense, does subject believe each involved the subject behaving more violently or aggressively than before?"

In responding to the teenage rape, Kohlhepp replied, "No. I pulled back and got shy again." But regarding the Superbike murders, he wrote, "Yes. Went looking for conflict and killing was only tool in the toolbox." And after his final murder: "Yes. After Charlie and while I held Kala, I was mentally setting myself up for next altercation."

Conditioned as he was by the lack of affection from his parents that helped create both his low self-esteem and his narcissistic personality disorder, with no siblings to rely on and with no chance to develop normal adolescent and young adult relationships with others, whenever Kohlhepp thought people were trying to take advantage of him, he had to get the jump on them.

Whatever Kohlhepp accomplished, it never seemed to impress either of his parents or earn him their respect, despite what Reggie later told interviewers. Like David Berkowitz, he came to feel he just wasn't wanted. And as was true with Berkowitz, much of his violence was the result of displaced anger.

When he called his mother after his arrest, Reggie's most direct question was the one that also said the most about her own self-involvement: "How can you do these things if you love me?"

"Because I messed up," he replied. "I'm sorry."

"Okay," Reggie responded.

"I love you," he declared.

"Okay" was all she could answer.

Talking with correspondent David Begnaud on the CBS television program *48 Hours*, she pointed out, "There was a lot of time between the first one and the others. And I know that doesn't mean much to the families, and I'm sorry. But he wasn't a serial killer."

Later she tried to explain the murders, saying, "They embarrassed him. And you know, anybody, I don't care who you are, whether you have a temper or you don't have a temper, nobody wants to be embarrassed. And that's hard to work out from." There is something both poignant and pathetic about a mother trying to comprehend the incomprehensible—her son's life as a murderer—and, I suspect, trying desperately to consider whether she had any role in that.

During his first interrogation session at the Spartanburg County detention center, Kohlhepp said, "I haven't been around my mom in years. I've tried. We had fallen apart."

She passed away shortly after that, on April 23, 2017.

"I miss my dog," he wrote to Maria. "I don't really miss my mother." Dogs offer unconditional and nonjudgmental love. Mothers don't always.

Ultimately, though, Kohlhepp recognized that no one else—neither Reggie nor Bill Sampsell nor Carl Kohlhepp nor the Superbike victims—was responsible for what he did.

Of all the repeat killers I've encountered, Kohlhepp provided us one of the most insightful, succinct, and accurate evaluations of his own psyche.

> *I had a very confusing, violent childhood with family that didn't want me, but didn't want anyone else to have me, either. They hated if I did something* [better] *than what they did. They were not proud of my college degree, pilot's license, or building*

a company. All they had was [a] hand out and snide comments. I have never felt at home with any of my family. I pushed so hard to prove I could become something, prove them wrong, and for the most part I did. But the residual parts of a bad childhood and then prison for so long just wouldn't let me walk away from violence and slights.

BUT IN SPITE OF THAT, HE UNDERSTOOD—OR AT LEAST WAS WILLING TO ADMIT—FAR more than most violent predators that his unfortunate background had not robbed him of free agency when he concluded, "My crimes are things I can control. My trigger finger, my call. No one made me do anything."

EPILOGUE

A KILLER'S CHOICE

On the afternoon of June 2, 1985, thirty-nine-year-old Leonard Lake returned to a hardware store in South San Francisco to pay for a vise that his friend Charles Ng had shoplifted earlier that day. Ng, twenty-four, shared Lake's remote cabin about a hundred miles away near Wilseyville, in Calaveras County. What caused this minor stab of conscience from Lake, we can't be sure. But it didn't go very well.

When the store clerk asked for ID, Lake didn't match the photo on the driver's license he handed over, in the name of Robin Stapley. The suspicious clerk called the police, who had already been alerted to the earlier shoplifting incident and arrived before Lake could get away. When the officers searched the trunk of his car, they found a .22-caliber handgun with a prohibited silencer, which was sufficient grounds for arrest.

The license plate on the car was registered to Lake, whom police had finally identified through his fingerprints, but when police ran the VIN number, the car belonged to a Paul Cosner, who had disappeared the previous November. The Robin Stapley on the driver's license had been reported missing by his family several weeks earlier. Under questioning by detectives, Lake gave up Ng, saying he was the one who had stolen the vise and Lake was just trying to make things right. Then he asked for a glass of water. He took two pills with it, then wrote a short note he said was for his family. The pills turned out to be cyanide tablets that Lake had sewn into his shirt collar. He died four days later without regaining consciousness.

People don't generally commit suicide over shoplifting charges or even over possessing illegal firearms, fake driver's licenses, or even potentially stolen cars. Police knew there had to be something else going on.

And indeed, there was. When detectives searched Lake's property in Wilseyville, they discovered a makeshift dungeon behind the cabin and a burial site with burned and crushed bone fragments that corresponded to at least eleven bodies. Two other buried bodies turned out to be those of Shapley and Lonnie Bond, one of Lake and Ng's neighbors. Two buried five-gallon buckets contained IDs and personal items of about twenty-five others, as well as Lake's handwritten journals for the past two years and two videotapes documenting the sexual assault and sadistic torture of two women, Brenda O'Connor and Deborah Dubs. Ng, meanwhile, had gone on the lam.

When these tapes and other evidence of Lake's and Ng's crimes reached us at Quantico, they were among the most depraved and sickening things I had seen in all my years in criminal justice. The only others that equaled this savagery for pleasure were Lawrence Bittaker and Roy Norris, who had met in prison and when paroled, decided to abduct, rape, torture, and kill one girl for each teenage year, thirteen through nineteen. They had already done this to five young women when one managed to escape and went to the police. Not quite as sophisticated as Lake and Ng, they only audiotaped their rape and torture sessions. Bittaker remains on death row in San Quentin, nearly forty years after his conviction. *What's wrong with that picture?* Norris took a plea in exchange for testifying against Bittaker: life with the possibility of parole. If parole is ever granted, it will be one of the greatest miscarriages of justice in California state history, and that's saying a lot.

During the filming of *The Silence of the Lambs* at the FBI Academy, I played one of those rape/torture/murder tapes for Scott Glenn, the fine actor who portrayed Jack Crawford, the profiler supposedly based on me. He was a sensitive, compassionate, and intuitive guy, the father of two daughters, a person who believed in rehabilitation and fundamental goodness. I saw tears start to well up in Glenn's eyes as he listened to the tape. Afterward, still in a state of shock, he said to me, "I had no idea there were people out there who could do anything like this." He told me he could no longer oppose the death penalty.

Charles Ng was located in Calgary, Canada, where his sister lived, about a month after Lake's suicide, when he shot a security guard in the hand while resisting arrest for shoplifting a can of salmon from a department store. He was tried and sentenced to four and a half years in prison, all the while fighting extradition back to the United States. He was finally sent back to California and indicted on twelve counts of first-degree murder.

Through a long series of legal maneuvers and complaints filed against prisons that held him and judges who ruled on his case, as well as churning through about ten attorneys, Ng was able to postpone his trial until the late 1990s. By that time, I had retired from the bureau and his then-current lawyer contacted me by phone.

He said he'd like to hire me to consult for the defense and started to give me the background, but I interrupted him and said, "I know the case." It had been presented to my unit and me back in the late 1980s by a California detective going through the National Academy program. The attorney went on to explain his theory of the case and the way he intended to present his defense: that Lake was the dominant and principal offender, and Ng, fifteen years younger and more mild-mannered, was essentially conditioned and coerced into following him and participating in the torture/homicides.

I said that from what I recalled, it didn't appear to be any sort of master-slave relationship.

The attorney replied something to the effect that if I saw all of the evidence and supporting material, I would see for myself that Ng had not participated willingly. I told him my hourly rate and gave him my standard pitch that I would approach the material with an open mind, and he could either use or not use my conclusions as he wished. He agreed and said that the state of California would pay for my consultation.

Within a week, I received a box that included crime scene photos, investigative reports, and all of the other materials I would need to evaluate and analyze the case, including the set of videotapes.

I looked into both Lake's and Ng's backgrounds.

Lake was six when his parents separated, and he and his sisters moved in with their grandmother. He had an early obsession with pornography; he was fond of photographing his sisters naked and bribed them into performing various sex acts with him. He was also fond of watching mice die by dissolving them in chemical solutions. He was discharged from the Marine Corps for psychological reasons and made sadomasochistic porno movies while living in a California commune. Ng was born in Hong Kong to wealthy Chinese parents. As a boy, he frequently got into trouble and was harshly disciplined with beatings by his rigid businessman father. He was kicked out of several boarding schools in Europe, then came to the United States and also joined the Marines, but faced court-martial for weapons theft in Hawaii. After an escape from the brig, he served eighteen months at Fort Leavenworth in Kansas before being dishonorably discharged. He then reconnected with Lake, whom he had met through a gamer magazine three years earlier.

None of this surprised me. I expected they'd come from dysfunctional backgrounds, and as the older of the pair, Lake would likely be the dominant one. But I was finding nothing that suggested Ng was under his thrall. In fact, on one of the stomach-turning tapes I reviewed, Ng tells their terrified victim, "You can cry and stuff, like the rest of them, but it won't do any good. We are pretty coldhearted, so to speak."

Going through the file, I even came to believe that when Lake went to the hardware store to pay for the stolen vise—which was to replace a broken one he and Ng had repurposed into a torture device—he was trying to compensate for his companion, who was a chronic thief, so as to smooth things over and not have a theft complaint put the police on their trail.

After I'd put in about twenty harrowing hours on the case, I figured I'd better call Ng's defense attorney and tell him where the evidence was leading me. I advised him that from everything I had seen so far, his client was a willing participant, and that I saw no evidence of coercion or even coaching from the late Mr. Lake. Ng on his own tormented his victims by cutting their underwear off them with a knife while Lake filmed the abuse.

Keep going, the attorney urged me. If I continued to go through more of the evidence, I would start to see what he was talking about. While he was no angel, the worst that could be said about Charles Ng was that he was an unwilling and compliant victim himself.

Reluctantly I agreed to proceed. But after about another ten hours of work that only confirmed what a hideous monster Ng was, I decided I was not going to waste any more of my time or the state's money on what, to me, was a ridiculous and unsupportable assertion.

I called the attorney and gave him the bad news. It was clear to me that his client willfully participated in these torture/homicides, and nothing I had seen had changed my opinion. The attorney was not pleased. I would say he was downright pissed off, and I had to remind him that I had told him at the outset that buying my time could not influence my opinion; only evidence could do that.

About a year later, I was not surprised to learn that the defense was able to find "experts" willing to testify on Ng's behalf. At his trial, whose venue was moved down south to Orange County to avoid prejudicial publicity, a psychiatrist testified that Ng had a dependent personality disorder. But under the prosecution's cross-examination, he admitted he had not viewed the tapes that I had. A psychologist had seen the tapes, but he opined that Ng's clearly sadistic behavior was merely a mirroring of Lake's behavior to please him.

Ng also decided to take the witness stand in his own defense, which afforded the prosecution to introduce even more evidence, including photographs of drawings he had made depicting the tortures he and Lake had committed, which he hung on the wall of his prison cell.

On February 11, 1999, Ng was convicted in eleven of the twelve counts of first-degree murder against him, with the jury deadlocking on the twelfth count. On June 30, Judge John Ryan accepted the jury's recommendation of death, stating, "Mr. Ng was not under any duress, nor does the evidence support that he was under the domination of Leonard Lake." As of this writing, Ng remains on death row in San Quentin, not far from Lawrence Bittaker.

WHEN I BEGAN MY OWN RESEARCH WITH THE FBI, I BELIEVED I WOULD FIND THAT almost all violent offenders were criminally insane because the cases we were receiving for analysis were so extreme in their perpetrators' violence against their victims. I thought, *There is no way that this degree of "overkill" or cruelty makes any sense.* After all, when we encounter crimes as heinous and disgusting as Ng's, one of the first questions we ask ourselves is *How could somebody do this to another human being?* These acts are unspeakable, often the works of highly disturbed minds. But the more we delve into the minds and personalities of these offenders and the better we are able to relate them to the crime scene evidence, the more we understand about the psychodynamics of the behavior.

As with all of the killers profiled here, nature versus nurture will always be the central tension in this debate about what could produce such monstrous and unnatural acts, but the exploration cannot end there. Inevitably the question eventually comes down to moral agency against inborn determinism. And that leads to a single word: *choice.*

Many killers would have it that their murderous actions were not their own choice at all, that killing was a fixed, nonnegotiable action for them. Yet nothing I have seen in all my years of criminal investigation leads me to accept that premise, except in the most extreme cases of mental illness.

The reality is that a bad background is not an excuse for murder; it never has been and never can be. I would not argue that Lake and Ng had the kind of nurturing, or lack thereof, that combined with their inherent natures to pave the way for their vicious crimes. As we've seen with both Joseph McGowan and Todd Kohlhepp, as well as with Edmund Kemper, David Berkowitz, and so many others, killers will look outside themselves to explain away their crimes.

Such was the case a few years after Ng's sentencing when forty-one-year-old John Allen Muhammad and seventeen-year-old Lee Boyd Malvo were arrested for the 2002 Beltway Sniper murders of ten and grievous injury of three more over a three-week period in Washington, D.C., Maryland, and Virginia. Muhammad refused to open up in custody, but Malvo talked freely. He was questioned by Detective June Boyle, a twenty-six-year veteran of the Fairfax County police department, who had investigated the murder of Linda Franklin, shot in the head by the sniper in a Home Depot parking garage while she and her husband, Ted, loaded shelving material into their car.

What everyone who witnessed the interrogation or listened to the audio recording afterward noted was how lighthearted, unconcerned, and unrepentant Malvo seemed. When Boyle asked him if he had shot Ms. Franklin, he matter-of-factly affirmed that he had. When she asked if he had seen through his rifle scope where she had been hit, according to Boyle's subsequent testimony at a pretrial hearing, Malvo "laughed and pointed to his head."

Boyle said Malvo seemed amused as he described the second Washington-area kill, of landscaper Sonny Buchanan. What struck Malvo as so funny was that after the victim was struck and toppled off the riding mower, it kept on going without him.

Dr. Stanton E. Samenow, a D.C.-area clinical and forensic psychologist who has been a hero of ours for a long time, was asked by the prosecutor to interview Malvo pretrial. With the late psychiatrist Dr. Samuel Yochelson, Samenow had cowritten *The Criminal Personality,* a landmark three-volume study based on extensive research the two conducted on violent criminals at St. Elizabeths mental hospital in Washington.

Samenow thoroughly reviewed Malvo's Jamaican upbringing, his father's noninvolvement, his mother's frequent absences and even more frequent physical punishment, and all the trouble he got into before he hooked up with the grifter and hustler Muhammad.

Malvo's defense was claiming that at the time he was suffering from what *DSM-4* classified as a "dissociative identity disorder, not otherwise specified." This meant that Malvo was so under Muhammad's influence that he was not himself.

Samenow told us, "This is not an exact quote, but one of the attorneys said something to the effect of: 'It was as though a pure river had been contaminated by a foul sewer; so was Lee Malvo poisoned, contaminated, and brainwashed by John Muhammad.' " Sound familiar?

In an interview from prison aired on NBC's *Today* show around the tenth anniversary of the D.C. Sniper crimes, Malvo said of Muhammad, who was executed for his crimes, "I couldn't say no. I had wanted that level of love and acceptance and consistency for all of my life and couldn't find it. And even if unconsciously, or even in moments of short reflection, I knew that it was wrong, I did not have the willpower to say no."

Yet ten years earlier, when Samenow asked him if he ever neglected or refused to do what Muhammad told him to do, the young man responded, "Oh, all the time."

In other words, he had a choice.

We study these people not as psychologists or sociologists, but as criminologists. We examine their backgrounds and upbringings to help us understand why they do what they do and how they go about it—to understand motivation and predict behavior—so we can apply it to our discipline of crime-solving and our mission of criminal justice. This means coming to grips with the question of why they make the choices to hurt and kill. Understanding how those choices are made, the pre- and post-offense behavior that goes into them, and the means by which those choices were carried out is the basis of criminal profiling.

Why? + How? = Who.

We are never going to get to the end of the rainbow in our study of human behavior, any more than we are ever going to eliminate crime. All we can do is to continue working at it, always trying to increase our understanding and knowledge.

The offenders we've examined in this book were all killers, yet they were all different. Every killer and predator represented many layers of subtle distinctions. The crime itself is the reflection of that difference, and generally points directly to motive. Yet we can say that all of these men had internal conflicts between grandiosity and inadequacy. All had a sense of personal entitlement that removed from them the obligation to heed society's laws and conventions. And all had the ability to make choices.

Someday neuroscience may be able to explain and pinpoint behavior to such a degree that we will be able to attribute a given thought to a specific morphological structure and electrochemical transmission within the brain. Yet even if that were to occur, would such a precise level of scientific reductionism and behavioral determinism wipe out the concept of personal responsibility? And if the answer to that question is yes, then what kind of society and moral universe would we be living in?

From the beginning of my law enforcement career, I can think of very few violent predators I or any of my colleagues interviewed that we would consider insane by the legal definition. They certainly were not normal, and most or all of them had some form of mental illness. But they knew right from wrong and the nature and consequences of their actions against others.

We often compare the cases of Susan Smith, who murdered her

three-year-old and fourteen-month-old sons in South Carolina in 1994, and Andrea Yates, who murdered her five children, aged six months to seven years, in Texas in 2001. Both sets of children died by drowning: the Smith boys when the car they were strapped into sank in a lake; the Yates children in their own bathtub.

Smith claimed her Mazda had been carjacked by an African American man with the boys in it. She went on national television to plead for their safe return. The police suspected her from the beginning, and her motive turned out to be an attempt to save her relationship with a wealthy man who had no room in his life for her children. Sadly, this motive is not uncommon in cases of parents killing children.

Yates, who had a long history of mental illness, postpartum depression, and attempted suicide and had been under psychiatric care, waited until her husband left for work, because she knew he would try to stop her. She drowned all five of her children, then called the police. Her motive was her belief that she was not a good mother, that Satan had therefore taken possession of her children, and that this was the only way she could save them from the fires of hell.

While both child murderers were arguably mentally ill, Susan Smith clearly knew the difference between right and wrong and made a decision that she perceived to be in her own best interests, even though it meant disposing of her children. Andrea Yates, on the other hand, was so delusional that she had no grasp of reality. While both cases are abysmally tragic, we would argue that only Smith's actions were evil. She made a considered choice. Yates was incapable of that.

Yates was not a predatory or serial killer. Had she not been caught, she would not have gone on to plan the murders of other family members or strangers to save them from hell. With the exception of very few outliers like blood-drinking Richard Trenton Chase and human-skin-wearing Edward Gein, who were so delusional that they couldn't discern reality, most killers know exactly what they're doing.

I believe all of the killers we've studied in this book had character defects that made them do what they did, rather than incapacitating psychoses. The people we're talking about wanted their victims to die, but they wanted to live. Within their own perverted value systems, that's pretty rational.

Joseph Kondro and Donald Harvey made deliberate choices involving complex planning. Todd Kohlhepp even wrote that he "made the choice to assault" the neighbor girl "and wreck everything." One might argue that McGowan and Kohlhepp, in his later killings, let emotion take over; that they were not actually making informed decisions. I would dispute this. As McGowan told me, as soon as he saw Joan D'Alessandro at the door, he had made the choice to *kill* her. As soon as he asked her to come downstairs with him, ostensibly so he could get the money for the Girl Scout cookies, he had made the choice as to *where* he was going to kill her. And as soon as he dashed her head to the floor, he had made the choice of *how* he was going to kill her. The same applies to Kohlhepp's murder of the Coxies and Charlie Carver. The legal concept of malice aforethought does not specify a time frame. It could be a year; it could be a split second. It is still a choice.

In the end, there is little that will ever feel complete when it comes to explaining the inexplicable mystery that lurks in the minds of killers. After a career spent studying these issues, I often find myself looking back to the idea expressed so nobly by Dr. Viktor Frankl, the Viennese psychiatrist, author, and Holocaust survivor whose book *Man's Search for Meaning* is one of the great moral and philosophical documents of our age. Frankl was able to find meaning even amidst the inhuman horrors of Auschwitz, where he lost his parents, brother, and pregnant wife. He took into account everything we are born with and everything that happens to us when he wrote, "Man is *not* fully conditioned and determined but rather determines himself whether he gives into conditions or stands up to them. In other words, man is ultimately self-determining."

All killers—unless they have a severe mental handicap or exist in a genuinely delusional state—are free to make their own choices. Yet even accepting that reality is not an end in itself. We must constantly try to expand our knowledge and increase our understanding of how and why the choice is made so we can better help law enforcement in identifying, catching, and putting these criminals away. That is why I started doing what I do, and that is why I continue facing the killers across the table.

ACKNOWLEDGMENTS

In one way or another, every book is a collaborative process, and we had plenty of help on this one. Special and heartfelt thanks go to:

Our editor, Matt Harper, for his constant enthusiasm and encouragement and the insightful guidance and clarity of vision that helped us tell this story as we wanted it; Anna Montague, Beth Silfin, Danielle Bartlett, Gera Lanzi, Kell Wilson, and the entire HarperCollins/William Morrow/Dey St. family.

Our ever-supportive and resourceful agent, Frank Weimann of Folio Literary Management.

Our UK editor, Tom Killingbeck, publicist Alison Menzies, and the team at William Collins in London.

Andrew Consovoy, former chairman of the New Jersey State Parole Board, and Robert Egles, former executive director.

Retired Tappan Zee High School teachers Robert Carrillo and Jack Meschino, and Jack's partner Paul Coletti.

Producer Trisha Sorrells Doyle and the MSNBC crew that worked with John on the Joseph Kondro and Donald Harvey interviews.

Committee Films producer-director Maria Awes, who has been incredibly generous in sharing her experiences, investigative research, communications, and analysis relating to her Investigation Discovery series, *Serial Killer: The Devil Unchained;* and the staff and crew at Committee Films, especially Jen Blanck and Bill Hurley for their meticulous review and fact-checking.

Forensic psychologist Stanton E. Samenow, Ph.D., from whose experience and vast knowledge we continue to benefit, especially here in relation to the D.C. Sniper case.

Mark's wife, Carolyn, who is our Mindhunters chief of staff and in-house counselor, among her many other talents and virtues.

John's daughter Lauren Douglas Skladany, for her careful readings and equally wise counsel.

And to Rosemarie D'Alessandro and her sons Michael and John, who have not only striven continually to keep Joan's legacy alive, but who have led an ongoing fight for justice and the safety of children everywhere. For that leadership, courage, and grace of spirit, this book is dedicated to her. Anyone interested in working with Rosemarie, taking up the cause of child safety to which she has dedicated her life, or contributing to the Joan Angela D'Alessandro Memorial Foundation can find details on the foundation's website: www.joansjoy.org.

ABOUT THE AUTHORS

JOHN DOUGLAS is a former FBI special agent, the bureau's criminal profiling pioneer, founding chief of the Investigative Support Unit at the FBI Academy in Quantico, Virginia, and one of the creators of the *Crime Classification Manual*. He has hunted some of the most notorious and sadistic criminals of our time, including the Trailside Killer in San Francisco, the Atlanta Child Murderer, the Tylenol Poisoner, the Unabomber, the man who hunted prostitutes for sport in the woods of Alaska, and Seattle's Green River killer, the case that nearly ended his own life. He holds a doctor of education degree, based on comparing methods of classifying violent crimes for law enforcement personnel. Today, he is a widely sought-after speaker and expert on criminal investigative analysis, having consulted on the JonBenet Ramsey murder, the civil case against O.J. Simpson, and the exoneration efforts for the West Memphis Three and Amanda Knox and Raffaele Sollecito. Douglas is the author, with Mark Olshaker, of seven previous books, including *Mindhunter*, the number one *New York Times* bestseller that is the basis for the hit Netflix series.

MARK OLSHAKER is a novelist, nonfiction author, and Emmy Award–winning filmmaker who has worked with John Douglas for many years, beginning with the PBS *Nova* Emmy-nominated documentary

Mind of a Serial Killer. He has written and produced documentaries across a wide range of subjects, including for the Peabody Award–winning PBS series *Building Big* and *Avoiding Armageddon*. Olshaker is the author of highly praised suspense novels such as *Einstein's Brain*, *Unnatural Causes*, and *The Edge*. In the other realm of life-threatening mysteries, he is coauthor with Dr. C. J. Peters of *Virus Hunter: Thirty Years of Battling Hot Viruses Around the World*, and with Dr. Michael Osterholm of *Deadliest Enemy: Our War Against Killer Germs*. His writing has appeared in the *New York Times*, the *Wall Street Journal*, *USA Today*, the *St. Louis Post-Dispatch*, and *Newsday*.

Both authors and their wives live in the Washington, D.C., area.